GCSE Practice Papers in Mathematics (Higher Level)

J.J. McCARTHY BEd (Hons)

DP PUBLICATIONS LIMITED
Aldine Place
London W12 8AW
Tel: (081) 746 0044

1990

ACKNOWLEDGEMENTS

I would like to thank J.S. Simmons of Stratford School for his help and advice at various stages in the preparation of the draft version of this booklet; I would also like to thank R.J. Chapman of DP Publications Limited for his continuous help and encouragement during the planning and writing of these papers.

A CIP catalogue record for this book is available from the British Library.

ISBN 1870941 20 9

First edition 1988
Reprinted with corrections 1989
Reprinted with amendments 1990

Typeset by
Alphaset
65A The Avenue, Southampton

Printed by
The Guernsey Press Co. Ltd.
Braye Road, Vale,
Guernsey, Channel Islands

Table of Contents

PREFACE

1. **Aim**

   The aim of this booklet is to provide *sufficient* GCSE Practice Papers to enable teachers, parents and pupils to know how well prepared the pupils are for their GCSE Mathematics examination at Higher Level.

2. **Need**

   The need was seen for a means of *testing* (both in the home and in the classroom) the *combination* of *breadth of knowledge* and the *speed needed* to convey that knowledge to the examiner.

3. **Approach**

   There are *two* sets of *one hour* Practice Papers for *each* topic or group of topics (Area and Volume: Matrices and Transformations etc.). Each set is followed by the relevant answers; teachers using the papers for class tests or for homework may wish to cut out the answers.

   In all, there are *ten* pairs of topic-based papers *plus* two *mixed* topics for mock examinations/final revision.

NOTES:-

(a) The GCSE Board's Syllabi, and their Papers, have all been closely studied and their main requirements incorporated.

(b) One hour was chosen as feasible in a *double period* at school and as a *reasonable time* to set aside for home self testing.

PART I

GCSE PRACTICE PAPERS

TIME ALLOWED
1 HOUR EACH PAPER

INTRODUCTION AND CONTENTS

Set aside an hour, *periodically*, to attempt each of the topic-based papers as you cover them in your studies. Check your answers *after* you have completed a paper.

In addition to the ten topic-based papers, there are two mixed example papers for final revision purposes. The first of the mixed papers has *short* questions and is designed to take you about 1½ hours. The second has a number of longer type questions for which around 12½ minutes should be allowed for each.

On the inside back cover is a formula sheet similar to one you will be provided with in your examination.

Paper

ALGEBRA AND GRAPHS

TIME 1 HOUR

1. (a) Solve the equation $5 - 3x = 2x - 10$.

(b)

$4x - 1$

$3 - 2x$

The perimeter of this rectangle is 8 cm.
Calculate x.

(c) Solve the simultaneous equations: $x - 2y = 4$, and $5x + y = 9$.

(d) Solve the quadratic equation $4x^2 - 7x - 2 = 0$.

2. (a) The equation of a straight line is $4x + 5y = 20$. What is its gradient?

(b) What is the equation of the straight line which passes through

$(1, -3)$ and $(2, -5)$?

(c) Taking values of x as ± 3, ± 2, ± 1, $\pm \frac{1}{2}$, draw a sketch of the graph of

$$y = \frac{1}{x^2}$$

(d) Use your graph to solve the equation

$$x = \frac{1}{x^2}$$

3.

x	−2	−1	0	1	2	3
y			−5			−5

(a) Copy and complete the table for $y = x^2 - 3x - 5$.

(b) On a sheet of graph paper choose suitable axes and draw the graph of the function.

(c) Draw a tangent to the graph at the point $x = 2$, and find the gradient at this point.

(d) Write down the equation of the line of symmetry of the curve, and state the co-ordinates of the point where it intersects the curve.

4. (a) $x = m^2 - n^2$. Calculate x when $m = 2\frac{1}{2}$ and $n = -1\frac{1}{2}$.

(b) Factorise the expressions: (i) $4x^2 + 8xy$ (ii) $3x^2 - 5x - 2$.

(c) Rearrange the formula $n = v - mx^2$, to make x the subject.

(d) In the formula $k^2 = 100 - m^2$, what are the largest and smallest numerical values m can have if k is a real number?

5. (a) y varies inversely as the square of x; if $y = 10$ when $x = 3$, find x when $y = 360$

(b) Simplify $-3(2x - 4) + 5(6 - 3x)$

(c) Factorise $ax + 3y - ay - 3x$

(d) Expand $(3 - 2x)^2$, and state its value when $x = -1$.

6. (a) Simplify $x^{2/3}$ when $x = 64$.

(b) The difference between two numbers is 6; their product is 16. Let the larger number be x, and form a quadratic equation in x. Hence find the two numbers.

(c) The gradient of a straight line is 3; it passes through (4,7). Calculate the intercept on the y-axis.

(d) The graph of a quadratic function intersects the x-axis at $x = 1$ and $x = 4$. Write down an equation of the form $y = ax^2 + bx + c$, where a, b, and c, are whole numbers, to satisfy the above.

AREA AND VOLUME
TIME 1 HOUR

1.

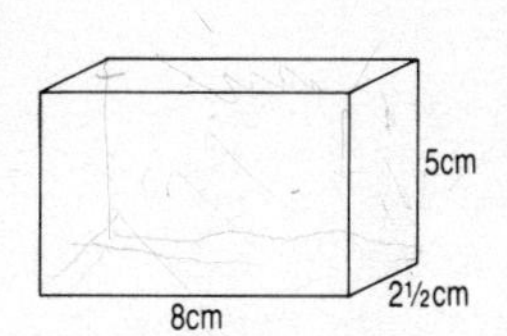

(a) What is the volume of the cuboid?

(b) Calculate the surface area of the cuboid.

(c) Another cuboid has measurements 16cm x 5cm x 10cm. What is the ratio of the surface area of the smaller cuboid to that of the larger?

(d) What is the least number of the original cuboids which must be stacked together to make a cube of side 40cm?

2. The diagram below shows a rectangular garden measuring 30m x 20m; there is a tiled path, 2m wide, all around the edge of the garden.

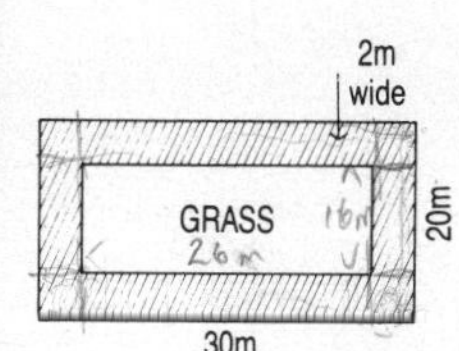

(a) Calculate the area of the grass portion of the garden.

(b) If the path is covered with square tiles 20cm x 20cm, calculate the number of tiles needed to cover the path.

(c) The owner decides to plant a circular flower bed of radius 2m in the garden; calculate the area of the flower bed, taking π as 3.14.

(d) In order to get the flower bed ready the owner had to dig the soil to a depth of 20cm; calculate in cm^3 the volume of soil she moved, giving your answer to 2 significant figures. Take π as 3.14.

3.

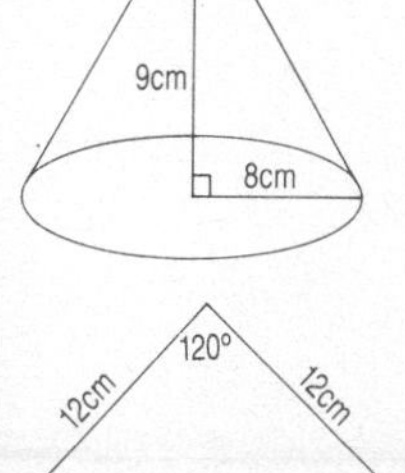

(a) Calculate the volume of the cone, giving your answer to the nearest whole number.

(b) Another cone has the same height, but its base radius is 6cm; what is the ratio of the volumes of the two cones?

(c) A cylinder of radius 8cm has the same volume as the cone; calculate the height of the cylinder.

(d) The sector of a circle (shown left) is to be folded to make a cone; calculate the radius of the base of this new cone. [You are advised not to substitute for π].

4.

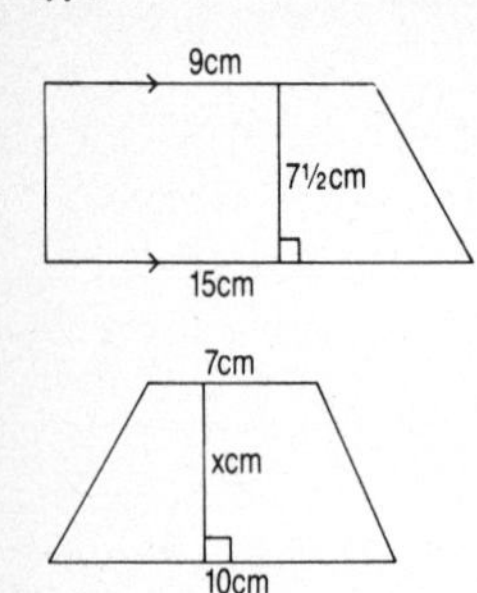

(a) Calculate the area of the trapezium.

(b) A similar trapezium has dimensions ⅓ of this one; state the relationship between the areas of the two trapeziums and use it to write down the area of the smaller one.

(c) What is the definition of a prism?

(d) The trapezium (shown left) is the cross-section of a block of timber 300cm long, with volume $0.03825m^3$. Calculate the value of x.

5. A factory makes spherical balls of radius 2cm; taking $\pi = 3.14$ calculate

(a) the volume of one ball, giving your answer to 1 decimal place.

(b) to the nearest whole number the area of paper required to wrap one ball.

Each ball is enclosed in a cubical box of edge 4cm, calculate

(c) the minimum area of cardboard needed to make one box.

(d) what percentage of the volume of each box a ball takes up.

6.

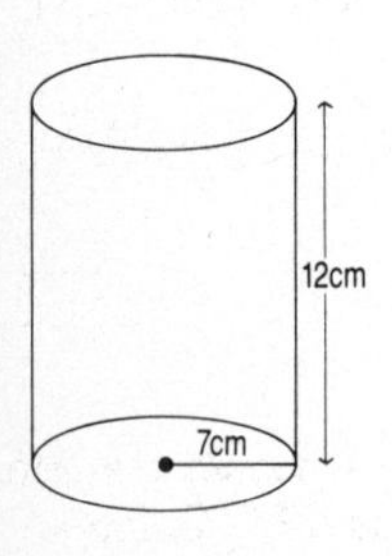

(a) Calculate the volume of the cylinder, giving your answer to 3 significant figures.

(b) Calculate the curved surface area of the cylinder, giving your answer to the nearest whole number.

(c) A box (cuboid) is made which just holds the cylinder; write down the volume of the box.

(d) The volume of a sphere is $70cm^3$; calculate its radius.

CIRCLE GEOMETRY (including intersecting chords and the length of an arc)

TIME 1 HOUR

Note: In this paper the point marked O *is the centre of the circle.*

1.

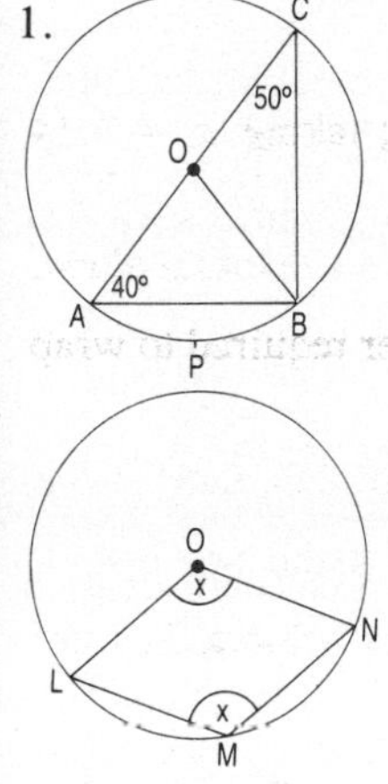

(i) In the diagram OA, OB, OC are radii; angle A = 40°, angle C = 50°.

(a) Calculate the size of angle AOB.

(b) Prove that A, O, C are in a straight line.

(c) If P is a point on the minor arc AB such that PB is parallel to AO, calculate angle PBO.

(ii) In this diagram L, M, N are three points on the circumference; if angle O = angle M calculate

(d) the value of x.

2.

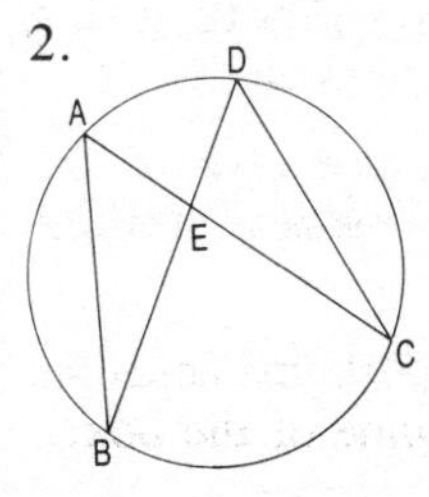

(i) In the diagram 2 chords AC and BD intersect at E

(a) Prove that triangles ABE and DCE are similar.

(b) Given that AE = 6cm, EC = 3cm, and BE = 4cm, calculate ED.

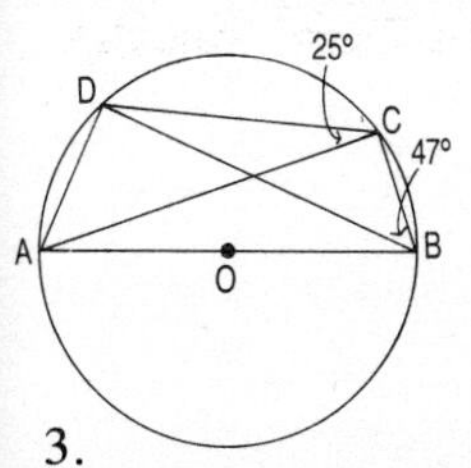

(ii) In the diagram AB is the diameter; angle ACD is 25°, and angle CBD is 47°. Calculate

(c) the size of angle ADC

(d) the size of angle CAB

3.

(i) In the diagram TA and TB are tangents to the circle.

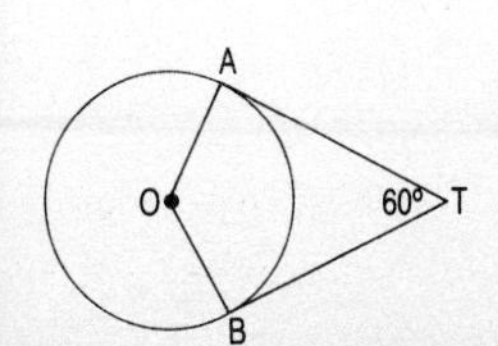

(a) What is the size of angle AOB?

(b) Explain why AOBT is a cyclic quadrilateral and state the centre of the circumscribing circle.

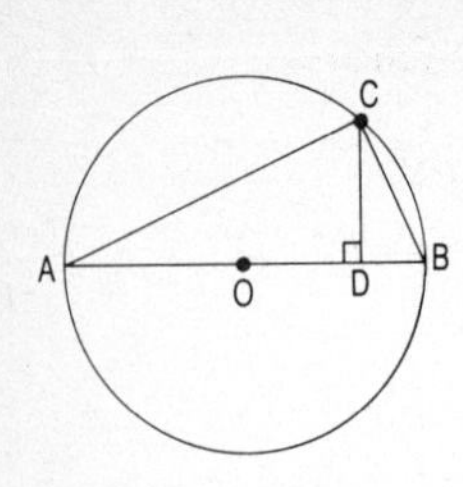

(ii) In this diagram AOB is a diameter and CD is perpendicular to AB.

(c) Prove that triangles CAD and CBD are similar.

(d) Given that AD = 6cm and DB = 4cm, calculate CD.

4.

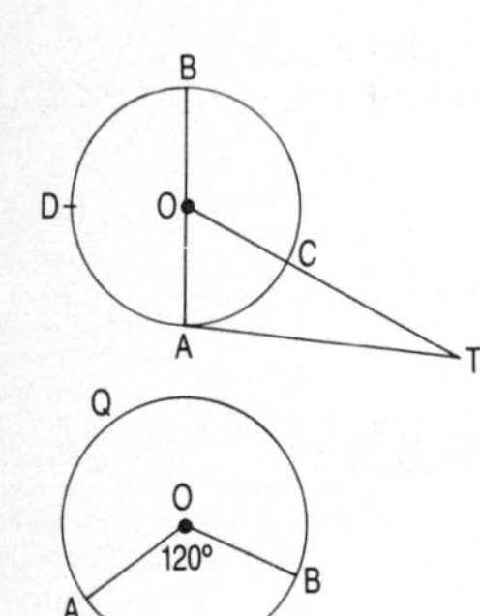

(i) In this diagram AOB is the diameter and TA is a tangent.

(a) If AB = 10cm and TA = 12cm, calculate OT.

(b) D is a point on the circumference such that AD = DB; calculate the length of AD.

(ii) Take the radius of the circle to be 6cm.

(c) Calculate the length of the arc APB – you may leave your answer in terms of π.

(d) Q is another point on the circumference such that the length of the arc AQ is

$$\frac{12\pi}{5}$$

Calculate the size of angle AOQ.

5.

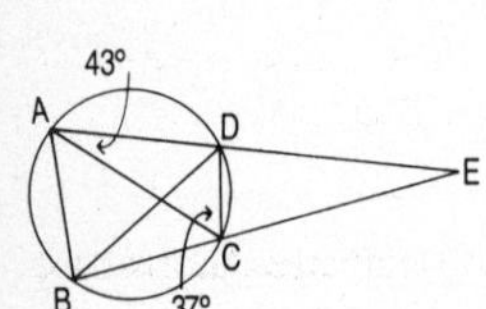

(i) ABCD is a cyclic quadrilateral with AD and BC produced meeting at E.

(a) Calculate the size of angle ABC.

(b) If also EC = ED, calculate the size of angle E.

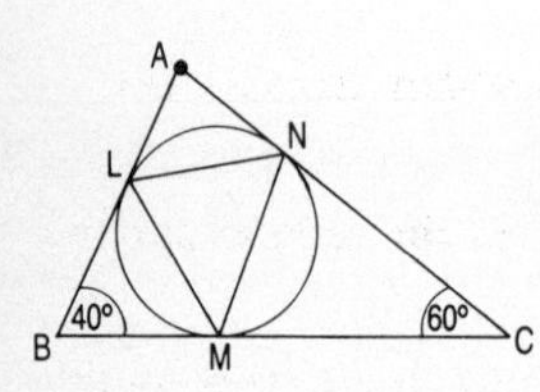

(ii) In the diagram the circle touches the 3 sides of $\triangle ABC$ at L, M and N.

(c) Calculate the size of each angle in triangle LMN.

"You cannot have a cyclic parallelogram."

(d) Do you agree?

6.

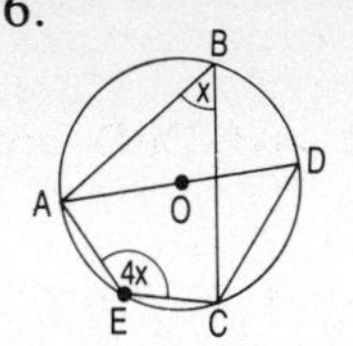

(i) Given that AOD is a diameter

(a) calculate the size of angle B

(b) if also BA = BC, calculate the size of angle BCD.

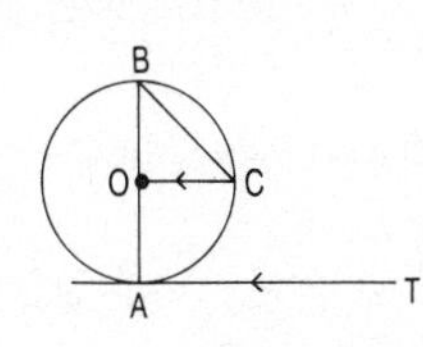

(ii) AB is a diameter of the circle and TA is a tangent at A; if TA is parallel to the radius CO,

(c) calculate the size of angle B.

(d) if also BC produced meets AT at G, write down the size of each angle of triangle CAG.

CONSTRUCTIONS (including bearings and scales)

TIME 1 HOUR

1. (a) Construct a triangle with sides 6cm, 8cm and 10cm.

   (b) Draw the circle that passes through each vertex.

The scale on a map is given as 1 : 25 000 000

   (c) how many kilometres does 1mm on the map represent?

   (d) how many square mm would represent an area of $2500km^2$?

2. A ship sails 720km from a port on a bearing 040°, and then turns 90°, to the right; it continues on its new course until it is 840km from its starting point.

   (a) Using a scale of 1cm = 120km, draw an accurate diagram of the first part of the journey.

   (b) Draw on your diagram the locus of points 840km from the ship's starting point.

   (c) Draw accurately the course of the ship for the second part of the journey.

   (d) On what bearing must it now sail in order to return to port?

3. A children's playing area is a rhombus with diagonals 16m and 12m.

(a) Make an accurate scale drawing of the playing area, using a scale of your own, state the scale you have used.

(b) Measure the acute angle of the rhombus.

(c) A certain group of children are forbidden to be more than 6m from either of the narrow ends of the playing area. Show clearly on your diagram the part of the area where they can play.

(d) Calculate the area of the playground *forbidden* to these children (to nearest whole number).

4. Construct the parallelogram ABCD with AB = 9cm, AD = 6cm and angle B = 120°. X is a moveable point inside the parallelogram, nearer to BC than AB; indicate by shading the region where X moves.

P is another moving point inside the parallelogram such angle DPA is 90°; draw the locus of P.

Q is a point on AB such that QB = 2cm. Draw a line from Q to DC that divides the parallelogram into two congruent trapezia.

5. AB is a straight line 8cm long;

(a) Construct on one side of AB the locus of the point C, where the area of triangle ABC is $12cm^2$.

(b) Mark on the locus you have drawn the point D such that DA = DB; join DA and DB.

(c) Bisect angles DAB and DBA; let the bisectors meet at K; Use K to construct the circle that touches the three sides of triangle DAB.

(d) Construct on the other side of AB the locus of points M such that angle AMB is 70°.

GEOMETRY OF RECTILINEAR FIGURES

TIME 1 HOUR

1.

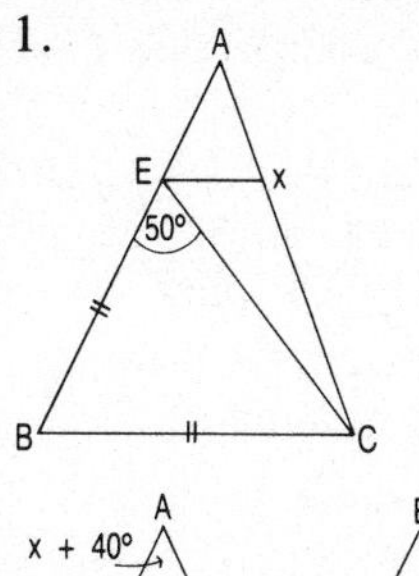

(i) In this triangle BE = BC and AB = AC. Angle BEC is 50°.

(a) What is the size of angle ECA?

(b) Given that EX is parallel to BC, calculate the size of angle XEC.

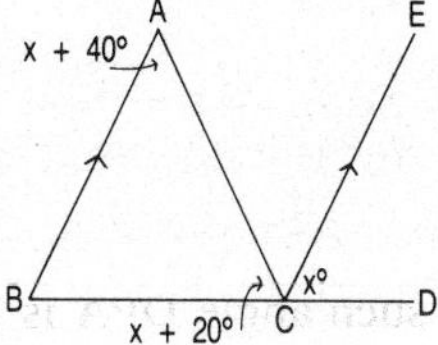

(ii) Given that EC is parallel to AB, calculate

(c) the value of x.

The interior angle of a regular polygon is 162°;

(d) how many sides has the polygon?

2.

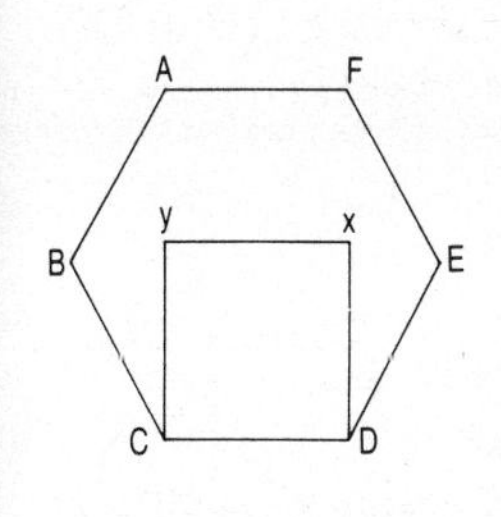

(a) 4 of the angles of a hexagon are 100°, 120°, 140°, and 170°; the other two angles are equal. Calculate the size of each.

The diagram shows a square standing on the side CD of a regular hexagon.

(b) How many lines of symmetry has the complete shape?

(c) What is the size of angle DEX?

(d) Calculate the size of each angle of triangle FXE.

3.

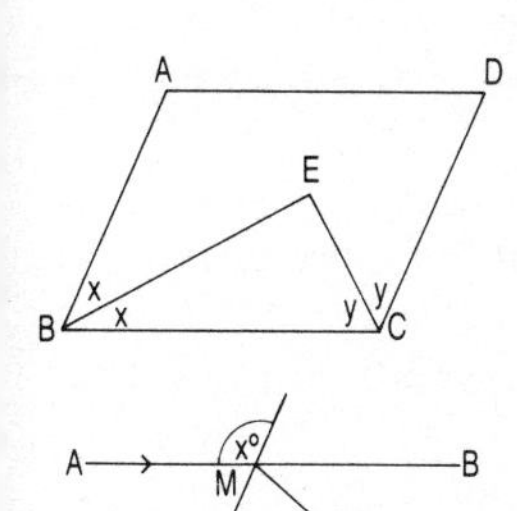

(i) ABCD is a parallelogram; the bisectors of angles B and C meet at E.

(a) Prove that angle E is 90°

(b) Given that angle A is 110°, state the value of y.

(ii) In the diagram AB is parallel to CD and EN = EM.

(c) Given that angle MED is 100°, calculate the value of x.

(d) Suppose the size of angle MED is unknown, and that EM produced bisects the angle marked x; calculate the size of angle MED then.

4. (a) Explain why each interior angle of a regular polygon cannot be 130°.

(b) ABCD is a parallelogram with AD produced to E so that AD = DE; if angle DBC is 30° and angle A is 130°, calculate the size of angle BCE.

(c) Which of these statements, if any, is true:

"Every rhombus is a square" or "Every square is a rhombus"?

(d) In a regular pentagon each line of symmetry divides the pentagon into 2 congruent trapezia. How does each line of symmetry divide a regular hexagon?

5.

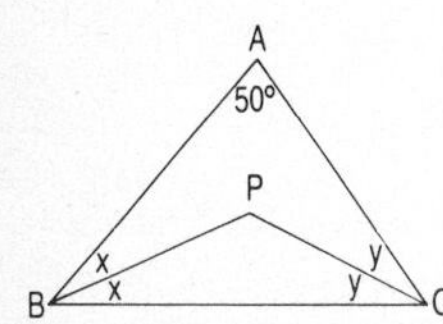

(i) In the triangle ABC, the bisectors of angles B and C meet at P; angle A is 50°.

(a) Calculate the size of angle BPC.

(b) If in general angle A = k°, express angle BPC in terms of k.

(ii) The ratio of the size of the interior angle to the exterior angle of a regular polygon is 7 : 1.

(c) Calculate how many sides the polygon has.

(d) What is the sum of its internal angles in degrees?

6. The regular pentagon ABCDE is rotated 40° anticlockwise to A′B′C′D′E′.

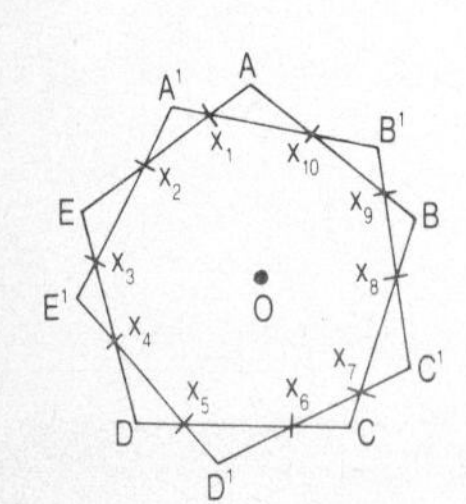

(a) Write down the size of angle A′.

(b) Prove that triangles OAA′ and OBB′ are congruent.

(c) Write down the size of each angle of triangle OAA′.

(d) What is the sum of the degrees of the decagon $X_1X_2. \dots \dots \dots .X_{10}$?

MATRICES AND TRANSFORMATIONS

TIME 1 HOUR

1. (a) On a piece of graph paper plot the points A(1,1), B(3,1) and C(1,3). Join the points to make a triangle and label it P.

   (b) Draw an enlargement of P with scale factor 3 and centre (0,0); label it Q. What is the ratio of the area of P to the area of Q?

   (c) Write down the matrix for the transformation in (b).

   (d) What is the inverse of the matrix you have used for (c)?

2.

$$M = \begin{pmatrix} -1 & 0 \\ 0 & -1 \end{pmatrix} \qquad N = \begin{pmatrix} 0 & -1 \\ -1 & 0 \end{pmatrix}$$

   (a) Calculate M − N

   (b) What transformation does matrix M represent?

   (c) What transformation does matrix N represent? What is the inverse of N?

   (d) Calculate MN and state the transformation the product represents.

3.

$$K = \begin{pmatrix} a & 2 \\ 3 & 4 \end{pmatrix}$$

   (a) For what value of a is the determinant of K equal to 2?

   (b) Calculate K^2, using the value of a you have found.

   (c) By inspection or otherwise complete the blank matrix:

$$\begin{pmatrix} & \\ & \end{pmatrix} \begin{pmatrix} a & 2 \\ 3 & 4 \end{pmatrix} = \begin{pmatrix} 3 & 4 \\ a & 2 \end{pmatrix}$$

(d) $N = \begin{pmatrix} x & 1 \\ 9 & x \end{pmatrix}$

If N is a singular matrix, calculate x.
(NB. A singular matrix has determinant 0.)

If a singular matrix is applied to the co-ordinates of a shape, what happens to the shape?

4. $Q = \begin{pmatrix} -2 \\ 3 \end{pmatrix}$, $R = \begin{pmatrix} 4 \\ 1 \end{pmatrix}$

(a) Referring to the triangle in question 1, write down the co-ordinates of P under the translation Q.

(b) $m(Q) + n(R) = \begin{pmatrix} -4 \\ 6 \end{pmatrix}$. Calculate m and n.

(c) S = (d, 5). Calculate QS.

(d) Show that (Q + R)S is singular.

5. The co-ordinates of a triangle are X(4,2), Y(8,2) and Z(6,6)

(a) Reflect the triangle in the y-axis, labelling the co-ordinates X′, Y′ and Z′ respectively.

(b) What matrix corresponds to the transformation in (a)?

(c) If the diagrams in (a) and (b) were unlabelled, the original triangle could have been mapped on to the second triangle by a different transformation. Describe fully what this transformation could have been.

(d) Calculate the area of triangle XYZ, and state what the area would have been under an enlargement with scale factor ½.

6. (a) $A = \begin{pmatrix} 2 & 7 \\ 1 & 4 \end{pmatrix}$ Find A^{-1}

(b) $\begin{matrix} \text{TOM} \\ \text{MARY} \end{matrix} \begin{pmatrix} £3 & £2 \\ £5 & £1 \end{pmatrix}$

The matrix shows the pocket money of two children for Saturday and Sunday.
What matrix can be used to pre-multiply the given matrix, so that Tom's money is halved and Mary's is doubled?

$$\begin{pmatrix} & \\ & \end{pmatrix} \begin{pmatrix} 3 & 2 \\ 5 & 1 \end{pmatrix} = \begin{pmatrix} 1\tfrac{1}{2} & 1 \\ 10 & 2 \end{pmatrix}$$

(c) $P = \begin{pmatrix} x & 3 \\ 2 & x \end{pmatrix}$; if the determinant of P is 3, calculate x.

(d) When a matrix is used to transform a shape, what relationship between the shape and its image does the determinant of the matrix give us?

NUMBERS AND FINANCE

TIME 1 HOUR

1. (a) 12% of a sum of money is £87; what is the sum of money?

   (b) The price of an article was reduced by 8½%; a buyer saved £3.40 because of this. What was the price of this article?

   (c) A Building Society pays 8¾% per annum simple interest. Calculate the interest on an investment of £10,000 for two years.

   (d) Mr. Lewis invested £2,300 with Happy Finance who pay interest annually; at the end of one year his interest was £184. What rate of interest did Happy Finance pay?

2. (a) Write down two prime numbers between 40 and 50.

   (b) A child's toy-set has three small bells which chime at 3 minute, 5 minute and 7 minute intervals respectively. They all chimed at 12 noon on a particular day; what was the last time prior to 12 noon of that day that they chimed together?

   (c) Express 243 as a power of 3.

   (d) Express 216 as the product of prime powers and state which root of 216 can be deduced from your expression. Give the root.

3. (a) Give an intelligent estimate of $\frac{44.9 \text{ x } 29.3}{136}$, making clear your method. (No credit for calculator answers.)

   (b) Express $\frac{0.16}{3.2}$ in standard form.

   (c) $(729)^x = 81$. Find x. (Hint: write 729 as a power of 9).

   (d) Taking 5 miles = 8km, express 1 mile in cm, giving your answer in standard form.

4. (a) What is the difference between the largest and the smallest share when £760 is divided in the ratio 5 : 6 : 8?

   (b) Twelve workers can do a job in 5 days; how many extra workers will have to be taken on if the job must be completed in 4 days?

   (c) The crowd at a football match was given as 45,500 to the nearest 100; state the possible limits of the crowd.

   (d) Simplify: $(1\frac{1}{2} + \frac{2}{3}) \div \frac{5}{6}$ (no credit will be given for answers in decimals).

5. (a) Taking £1 = 203 Spanish pesetas convert 7,000 ptas to sterling.

(b) On a journey of 350 miles a driver covers the first 100 miles at 40 m.p.h., and the remainder at 62½ m.p.h. Calculate her average speed for the whole journey.

(c) Arrange in descending order: 71%, $^5/_7$, $(0.84)^2$.

(d) Show by an example of your own why the following statement is false:

"If you increase a number by 10% and then decrease the result by 10%, you will get the original number".

6. A new suite costs £350 for cash; a Hire Purchase scheme is available, under which the suite will cost £392. Under this agreement a purchaser will pay a deposit of 10% (of the cash value), followed by 7 equal instalments.

(a) How much will each instalment be?

(b) Express the difference between the cash price and the H.P. price as a percentage of the cash price.

(c) If the cash price represents a 25% profit for the store, calculate how much the suite cost the store.

(d) The store spent £84,000 on suites at the price in part (c); what percentage of this stock must it sell at the cash price in order to 'break-even'? (i.e. not make a profit and not make a loss.)

PYTHAGORAS AND TRIGONOMETRY

TIME 1 HOUR

1.

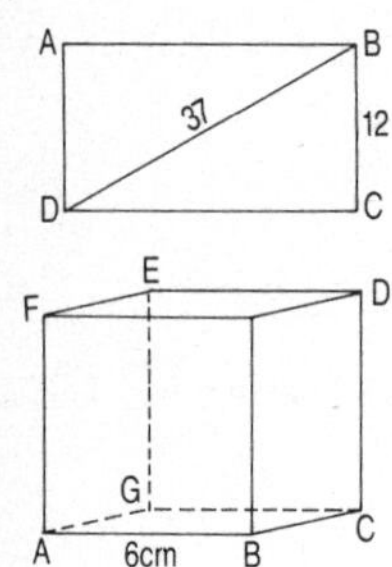

ABCD is a rectangle.

(a) Calculate the length of DC.

(b) Calculate the size of angle ADB.

The sketch is of a cube of size 6cm.

(c) Calculate the length of the diagonal AD (you are advised to make a sketch of triangle ACD).

(d) Calculate the size of angle DAC.

2.

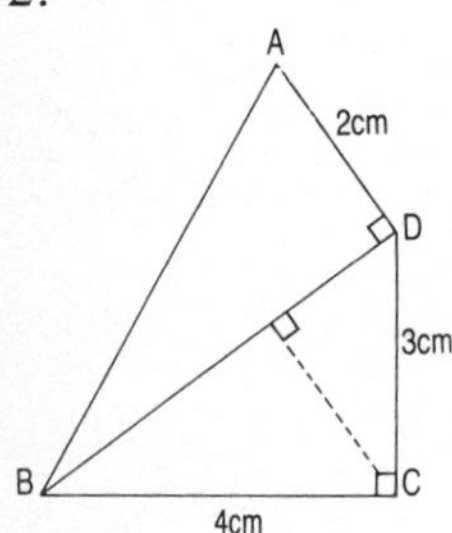

ABCD is a quadrilateral, with angle C = 90°, and angle ADB = 90°.

(a) Calculate the size of angle DBC.

(b) Calculate the length of the perpendicular from C to DB.

(c) Calculate the size of angle ABC.

(d) Calculate the length of the perpendicular from A to BC.

3. (i) To calculate the height of a tree Claire walked 60m from the base and found that the angle of elevation of the top of the tree was 18.4°.

(a) Calculate the height of the tree.

Claire's friend Paul is standing at a point from which the angle of elevation is half of what it is for Claire.

(b) How far is he from the tree?

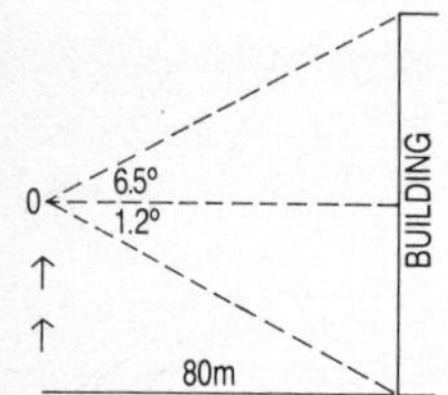

(ii) Peter walks 80m back from the base of a building and finds that the angle of depression of the base of the building is 1.2°, and the angle of elevation of the top is 6.5°. Calculate the height of the building. (Sketch to left).

4.

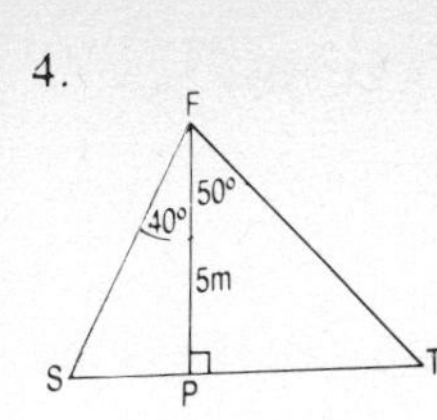

(i) To hold a vertical flagpole FP, 5m high, in position two ropes FS and FT are fastened to the ground at S and T; the points S, P and T are collinear.

(a) Calculate the distance SP.

(b) Calculate the length of the rope FT.

(ii) The sides of the triangle are 11, 60, and 61.

(c) Prove that the triangle is right angled.

(d) Calculate the length of the perpendicular from the right-angled vertex on to the hypoteneuse.

5. (a) If 4 Sin x = 3, calculate x.

(b) A road slopes up at 10°; how far will you rise if you walk 700m along the road.

(c) Given the Sin x = 15/17, calculate Cos x and Tan x. (No credit will be given for answers in decimals.)

(d) Simplify $\dfrac{\text{Sin } 30 \times \text{Sin } 40}{\text{Sin } 50 \times \text{Sin } 60}$

6. A boat travels at 12km/h due East from the base of a cliff, 300m high, until the angle of elevation of the top of the cliff is 14°. Calculate

(a) how far the boat is from its starting point.

(b) how long, to the nearest minute, it took the boat to reach its present position.

(c) An observer on the cliff-top, directly above the point of departure, moves 400m North. Calculate the angle of depression of the boat from her new position. (Assume the cliff is running North – South.)

PROBABILITY AND STATISTICS

TIME 1 HOUR

1. "90% of our girls and 80% of our boys pass Literature – which is an average of 85%" – Headmaster.

Using figures of your own, give an example to show

(a) that this claim could be true, and an example to show

(b) how this claim could be false.

The average weight of 9 footballers is 65kg;

(c) if the heaviest is omitted, the average drops by 2¾kg. Find the weight of the heaviest person.

(d) during the following year the weight of the heaviest player is expected to remain the same, but each of the other players is expected to 'put-on' 1½kg; what will be the average weight of the players a year from now?

2. (i) To find the probability of selecting 2 picture cards from a pack of cards (without replacement) a pupil submitted the following answer:

"The chance of the first card being a picture card is $\frac{12}{52} = \frac{3}{13}$,

so the chance of the second card being a picture card is $\frac{2}{12} = \frac{1}{6}$

hence the combined probability is $\frac{3}{13} \times \frac{1}{6} = \frac{1}{26}$".

(a) Explain the error in this answer.

(b) Calculate the correct probability of selecting 2 picture cards.

(ii) Two dice are thrown: what are the chances that

(c) the two results are different?

(d) that the difference between the two scores does not exceed 3?

3. The results of a Mathematics test were grouped as below:

MARKS	0-19	20-39	40-59	60-79	80-99
FREQ.	1	3	8	12	6

(a) What was the modal group of marks?

(b) What percentage of the candidates got 60 or more marks?

(c) Estimate the mean mark.

(d) If the results were represented by a pie chart, which group of marks would have a sector angle 96°?

4. A pupil tosses 3 coins:

(a) What are the chances she will get only 2 heads?

(b) She submitted the following correct answer to a question about the three coins: ANS: $1 - (\frac{1}{2})^3 = 1 - \frac{1}{8} = \frac{7}{8}$
What could the question have been?

On a pie chart representing the budget of a foreign city the angles representing Education, Scientific Research, Health, Others were 130°, 120°, 100°, and 10° respectively.

(c) If 8 million dollars were spent on Education, calculate to the nearest half million the budget of the city.

(d) The same information was shown on a bar chart; the height of the column for Education was 65mm. How high was the column representing Health?

5. From the set of numbers (1, 8, 1, 9, 3, 5, 1, 6)

(a) write down the mode and the median?

(b) if two numbers are chosen at random, what are the chances that both are less than the mean?

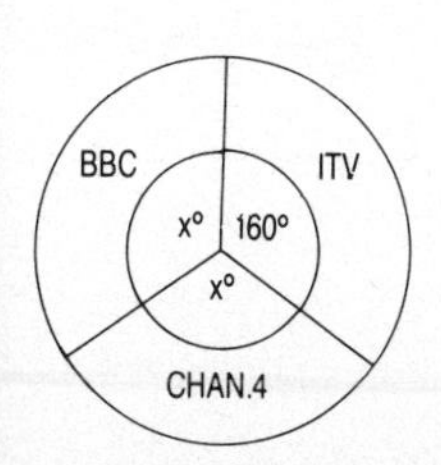

In a survey about TV preferences among 900 people BBC and Channel 4 were equally popular; the sector angle on a pie chart for ITV preferences was 160°.

(c) How many people preferred Channel 4?

(d) In a subsequent survey of the same people, it was found that ITV had lost 10% of its viewers to BBC; if no BBC viewer had changed preference, calculate the size of the sector angle which would now represent BBC viewers.

6. When you play a computer game for the first time in a series of games, your chances of winning are ¾ and of drawing $^1/_{20}$.

   (a) What are your chances of losing?

If in the course of a series of games, you win a game, the computer subtracts $^1/_{20}$ from your chances of winning a subsequent game; it adds an equal amount to your chances of drawing, and of losing.

   (b) Write down your chances of winning 2 games in the series.

   (c) If you win one game, what are your chances of drawing the next game?

   (d) What is the greatest number of games you could win in a series?

VECTORS

TIME 1 HOUR

1. $\underline{p} = \begin{pmatrix} 4 \\ 3 \end{pmatrix}$ $\underline{q} = \begin{pmatrix} -3 \\ 2 \end{pmatrix}$

Find

   (a) $\underline{p} - 3\,\underline{q}$

   (b) the modulus of q (leave your answer in root form).

If $\underline{r} = \begin{pmatrix} -5 \\ k \end{pmatrix}$ find

   (c) k if $\underline{r}$ is parallel to $\underline{p} + \underline{q}$

   (d) the modulus of $2\underline{r}$

2. ABCD is a rectangle with $\overrightarrow{AB} = \underline{a}$ and $\overrightarrow{BC} = \underline{b}$.

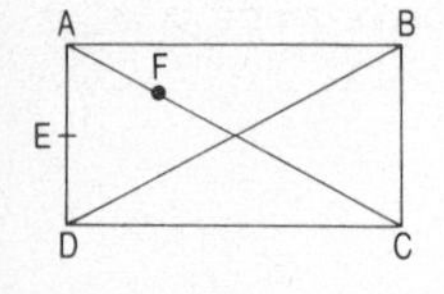

(a) Write down expressions for $\overrightarrow{AC}$ and for $\overrightarrow{BD}$ in terms of $\underline{a}$ and $\underline{b}$.

Given AE = ½AD and AF = ¼AC,

(b) find and simplify a vector expression for $\overrightarrow{EF}$

(c) show that EF is parallel to DB

(d) express in simplest form the ratio of the area of triangle AEF to triangle ABD.

3. $\vec{OA} = \begin{pmatrix} 2 \\ 4 \end{pmatrix}$ and $\vec{OB} = \begin{pmatrix} 4 \\ -2 \end{pmatrix}$

(a) If OA and OB are two sides of a parallelogram OBDA, find the position of vector $\vec{OD}$.

(b) Calculate the modulus of $\vec{OA}$ and of $\vec{OB}$, leaving your answers in root form.

(c) What do your answers in (b) tell you about the parallelogram?

(d) Calculate the length of each diagonal of the parallelogram and comment on the significance of your answer.

4. $\underline{a} = \begin{pmatrix} -2 \\ 1 \end{pmatrix}$ and $\underline{b} = \begin{pmatrix} -3 \\ x \end{pmatrix}$

(a) if $\underline{a}$ is parallel to $\underline{b}$, state the value of x.

(b) using your value of x, evaluate $6\underline{b} - 5\underline{a}$.

On a co-ordinate grid (with positive and negative axes) the position vector $\vec{OP} = \begin{pmatrix} 5 \\ 12 \end{pmatrix}$.

(c) write down the modulus of $\vec{OP}$

(d) using whole numbers only, write down 3 more vectors, $\vec{OR}$, $\vec{OS}$ and $\vec{OT}$, that have the same modulus as $\vec{OP}$.

5.

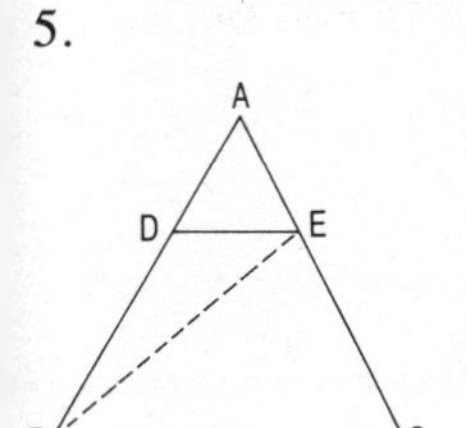

ABC is a triangle with $\vec{AB} = \underline{a}$ and $\vec{AC} = \underline{b}$. AD = ⅓AB.

(a) Write down $\vec{AD}$ and $\vec{BD}$ in terms of $\underline{a}$.

(b) Express $\vec{BC}$ in terms of $\underline{a}$ and $\underline{b}$.

(c) If $\vec{BE} = \frac{1}{3}\underline{b} - \underline{a}$, calculate DE in terms of $\underline{a}$ and $\underline{b}$, and deduce that DE is parallel to BC.

(d) State in simplest terms the ratio of the area of trapezium DECB to the area of triangle ABC.

6. (i) The position vectors $\vec{OA}$, $\vec{OB}$, $\vec{OC}$, relative to an origin O, are

$\begin{pmatrix}-4\\5\end{pmatrix}$, $\begin{pmatrix}0\\1\end{pmatrix}$, and $\begin{pmatrix}3\\-2\end{pmatrix}$ respectively.

(a) Show that A, B, and C are collinear.

(b) Find the position vector $\vec{OP}$, given that P is the point where BC intersects the x-axis. (No credit will be given for answers that rely on accurately drawn diagrams.)

(ii) Two non-zero vectors $\underline{p}$ and $\underline{q}$ are connected by the relationship $a\underline{p} + b\underline{q} = 0$, where a and b are scalars.

(c) what conclusion can you draw if $\underline{p}$ and $\underline{q}$ are not parallel?

(d) what conclusion can you draw if $\underline{p}$ and $\underline{q}$ are parallel, and both a and b are non-zero?

MIXED EXAMPLES 1

TIME 1½ HOURS (approx.)

1. Express in standard form:

(a) $(37000)^2$

(b) $(4 \times 10^6) \div (5 \times 10^8)$

2. A shopkeeper bought an article from a dealer for £8.60 and sold it for £9.89. Calculate

(a) the shopkeeper's profit as a percentage

(b) the cost of the article to the dealer if she made 25% profit by selling it at £8.60 to the shopkeeper.

3. (a) What is the gradient of the straight line $x + 2y = 6$?

(b) State the co-ordinates of the point where it intersects the line $y = x$.

4. Solve the quadratic equation $3x^2 - 7x - 6 = 0$. Use your solutions to make a rough sketch of $y = 3x^2 - 7x - 6$, showing where the graph intersects the axes.

5. (a) Express 210 as the product of prime numbers.

(b) What is the H.C.F. and the L.C.M. of 32 and 48?

(c) Simplify $27^{1/3} \times 64^{-2/3}$.

6. $A = \begin{pmatrix} 1 & 0 \\ 0 & -1 \end{pmatrix}$, $B = \begin{pmatrix} -1 & 0 \\ 0 & -1 \end{pmatrix}$

(a) Calculate AB

(b) What transformation does each matrix represent?

7. 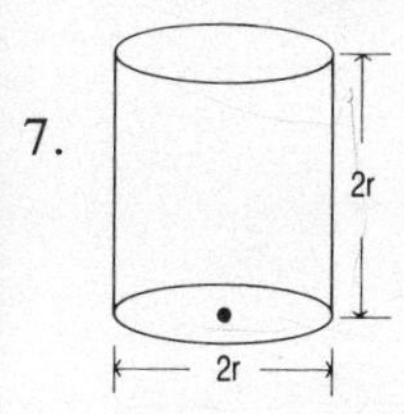

The diameter of the base of a cylinder is equal to its height. If the volume of the cylinder is 128π cm^3, calculate r.

8. Express a speed of 40km/h in metres/sec.

9. The radii of two similar cones are 4cm and 6cm. Express the ratio of their volumes in simplest form.

10. Sketch two different nets of a cube, neither of which has 4 squares in a row.

11. 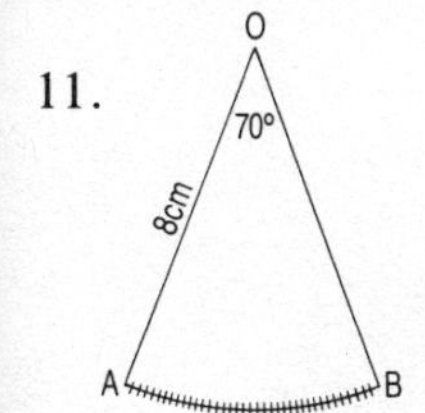

OAB is a sector of a circle, radius 8cm. If angle OAB = 70°, calculate

(a) the length of the arc AB

(b) the area of the sector.

12. Solve the equation $(x + 4)(x - 2) = 40$

13. A car travels from A to B at 40 m.p.h. and returns at 60 m.p.h. What is the average speed for the combined journey? [You may let the distance from A to B be a suitable number!]

14. Solve the simultaneous equations $3x - 2y = 8$; $4x = 7 + 2y$

15.

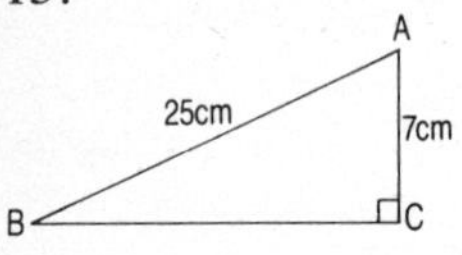

Calculate

(a) the size of angle A

(b) the area of the triangle.

16.

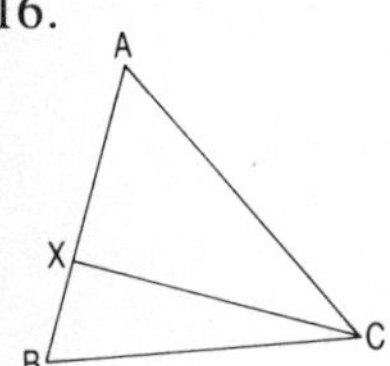

In the $\triangle ABC$, let $\overrightarrow{AB} = \underline{a}$, $\overrightarrow{AC} = \underline{b}$, and $AX : XB = 4 : 1$

Express in terms of $\underline{a}$ and $\underline{b}$

(i) $\overrightarrow{BC}$

(ii) $\overrightarrow{CX}$

17. The scale on a map is 1 : 100 000.

(a) How far apart, in kilometres, are two places which are represented on the map by a line 7mm?

(b) What is the area in km^2 of a lake represented on the map by $14cm^2$?

18. A ship travelling on a bearing of 280°, turns 90° to the left. What is its new bearing?

19. $\overrightarrow{OA} = \begin{pmatrix} 2 \\ 3 \end{pmatrix}$ $\overrightarrow{OB} = \begin{pmatrix} -3 \\ 2 \end{pmatrix}$.

Find

(i) $\overrightarrow{BA}$

(ii) angle OAB.

20.

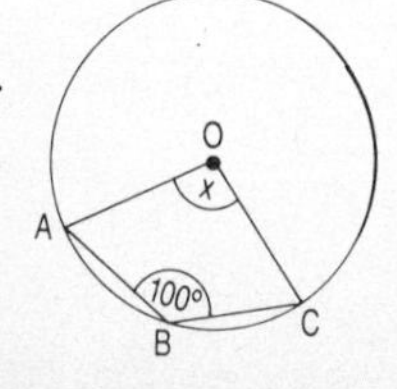

(a) Find the size of the angle marked x.

(b) If also AB = BC, calculate angle A.

21. For what values of x is:

(i) $4x + 1 < -9$?

(ii) $-1 - 4x > 99$?

(iii) $-8 < 3x + 1 \leqslant 37$, where x is an integer?

22. In a group of 20 pupils, they all play either football or hockey, or both; if 13 play football and 8 play hockey, how many play only 1 game?

23. Which of the following are Rational numbers:

(i) $\sqrt[3]{27}$

(ii) $(\pi)^2$

(iii) $-2\frac{1}{2}$

(iv) $\sqrt{3}$

(v) 3^{-2}?

24. Make a sketch of the graph of $y = \text{Cos } 2x$, taking values of x from 0° to 90°, at 15° intervals. Use 5cm for 1 unit on the y-axis.

25.

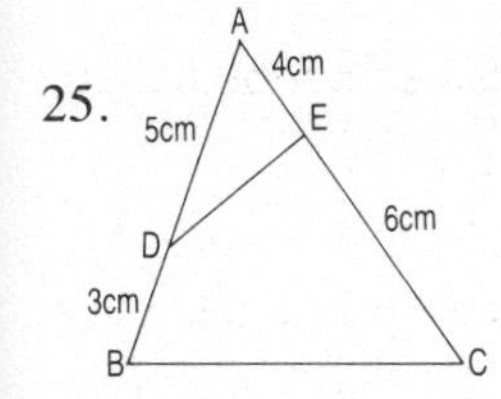

Express the area of triangle ADE as a fraction of the area of triangle ABC. [You are advised to draw each triangle on its own with its known measurements marked in.]

26. $f(x) = 1 - 9x^2$. Calculate

(a) $f(-2)$

(b) m if $f(m) = 0$.

27. Calculate the size of each interior angle of a regular 12-sided polygon.

28. What is the equation of the line which is parallel to $y = \frac{1}{4}x + 3$ and pass through $(2, -1\frac{1}{2})$?

29. $\underline{a} = \begin{pmatrix} 3 \\ 4 \end{pmatrix}$, $\underline{b} = \begin{pmatrix} 5 \\ 12 \end{pmatrix}$ What is the modulus of $\underline{a} + \underline{b}$?

30. A car is valued now at £8,000; if it depreciates in value by 10% annually, what will it be valued at in 3 years time?

MIXED EXAMPLES 2

NO TIME LIMIT

[Allow approximately 12½ minutes a question. Questions 11-15 may not all be on your syllabus.]

1.

t	0	1	2	3	4
h	12	12	10	6	0

The table gives the height (h) of a ball after t seconds.

(a) Plot the co-ordinates (t, h) and join them with a smooth curve.

(b) What was the height of the ball when t = 2½ seconds?

(c) Draw a tangent to the curve at the point where t = 3; estimate the gradient at this point.

(d) h and t are connected by the equation $h = at^2 + bt + c$, where a, b, and c are constants.

(i) Write down the value of c.

(ii) Using this value of c, substitute t = 1, and t = 2 in the equation to get two simultaneous equations. Solve the equations to find a and b.

2. (i) (a) The radius of a sphere is 4cm. Calculate its volume.

(b) Hence write down the volume of a sphere of radius 8cm.

(ii) 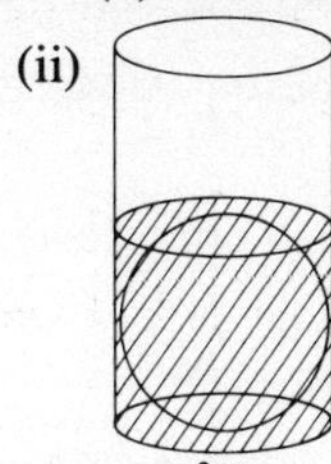

A sphere of radius 3cm is placed in a cylinder as shown; water is poured into the cylinder until the top of the sphere is just covered. Calculate the height of the water when the sphere is removed. [You are advised not to substitute for π]

3. Construct accurately the triangle ABC, AB = 8cm, AC = 11cm and BC = 13cm.

Draw the locus of the points which are equidistant from B and C, and the locus of the points equidistant from A and C. Hence draw the circumcircle of the triangle. Measure its radius.

Rotate the triangle 60° anticlockwise about the centre of the circle.

4. Draw the triangle ABC where A is (0, 2), B (4, 0) and C (2, −1).

(a) Rotate the triangle 180° about the point A.

(b) Reflect the triangle ABC in the x-axis, and give the matrix that represents this transformation.

(c) What vector of the form $\begin{pmatrix} a \\ b \end{pmatrix}$ translates the triangle ABC to (−4, −1) (0, −3) and (−2, −4)?

(d) What transformation does the matrix $\begin{pmatrix} 0 & -1 \\ -1 & 0 \end{pmatrix}$ represent?

5. (a) Express 729 as a power of 3.

(b) Simplify $64^{1/3} \times 32^{3/5}$.

(c) $a = 3 \times 10^5$ and $b = 4 \times 10^{-5}$, express ab in standard form.

(d) The population of a town decreased by 8% over the last 5 years. If it is now 46,000, what was it 5 years ago?

Because of a new industry the population is expected to rise again by 8% over the next 5 years. What should the population be in 5 years time?

6. (i) Given that O is the centre of the circle,

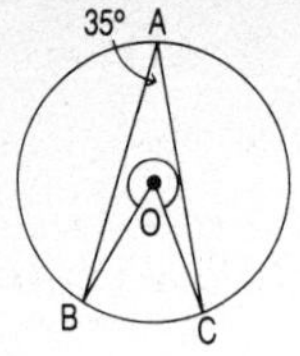

(a) Calculate the size of the reflex angle O, if A = 35°.

(b) State in general the relationship between the reflex angle O and angle A.

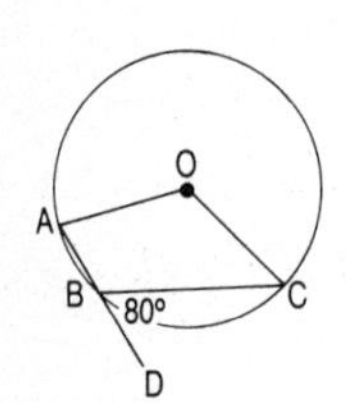

(ii) In the circle centre O, AB is produced to D; angle CBD = 80°

(c) Calculate the size of angle AOC.

(d) If also CB = CO, calculate the size of angle A.

7. (a)

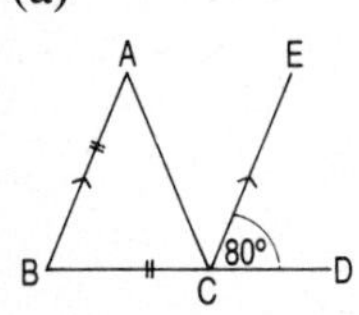

In the diagram BA = BC and AB is parallel to EC. Calculate the size of $A\hat{C}E$.

(b) The sum of the angles of a polygon is 1260°. How many sides has it?

(c)

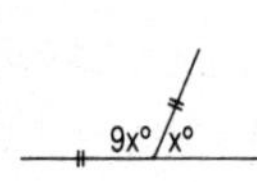

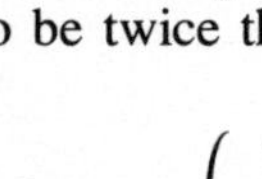

This is one vertex of a regular polygon, showing an exterior angle. How many sides has the shape?

(d) Is it true or false, in general, to say that if one polygon has twice as many sides as another, then the sum of its degrees will also be twice that of the other?

8. (a) Let $\underline{a} = \begin{pmatrix} 2 \\ -3 \end{pmatrix}$ and $\underline{b} = \begin{pmatrix} 5 \\ 2 \end{pmatrix}$

Calculate

(i) $3\underline{a} - \underline{b}$

(ii) the modulus of $\underline{a} - \underline{b}$

The position vectors $\overrightarrow{OA} = \begin{pmatrix} 2 \\ 2 \end{pmatrix}$ and $\overrightarrow{AB} = \begin{pmatrix} 4 \\ 5 \end{pmatrix}$

(b) Given that these are two adjacent sides of a parallelogram OABC, find the co-ordinates of C.

(c) What is the position vector of the point where the diagonals of the parallelogram intersect?

(d) Prove that the shape is not a rhombus.

9. (a) Jason sleeps 9 hours a day; what angle would represent this on a pie chart?

(b) A bag contains 8 blue buttons and 10 red ones. If two buttons are drawn at random, what are the chances the second button is blue?

AGE	5	6	7	8	9	10
FREQ.	3	2	5	3	7	10

The table above gives the results of a survey on children's ages.

(c) What was the mean age?

(d) If two children are chosen at random what are the chances that at least one is 10 years old?

10. (a) The diagonals of a rectangle intersect at 40°; if each diagonal is 10cm long, calculate the length of the sides of the rectangle.

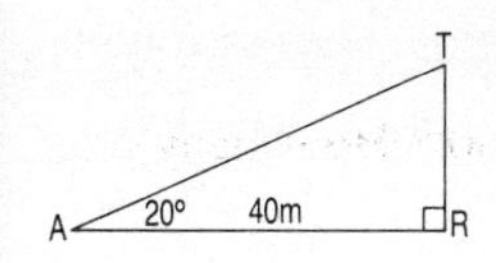

(b) The angle of elevation of the top of a tree from a point 40m from its base is 20°. What will be the angle of elevation if the observer moves 20m nearer to the tree?

(c) Given that tan B $= \frac{12}{35}$, calculate sin B giving your answer as a fraction.

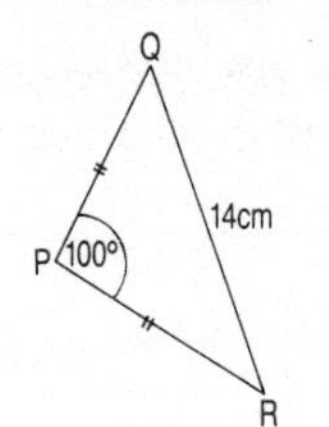

(d) PQR is an isosceles triangle with QR = 14cm, angle P = 100°. Calculate the length of PQ.

11. $f(x) = 2 - 3x$ and $g(x) = x^2 + 2$

(a) find (i) $f(-2)$ (ii) $gf(-2)$

(b) find a if $2f(a) = f(3a)$

(c) express $fg(x)$ in simplest form

(d) for what values of x is $f(x) = g(x)$?

12. A finance company pays interest 6-monthly on an investment of £80,000; if the annual rate is 9%, calculate how much the investment will amount to after 2 years. Give your answer to the nearest £100.

Suppose the £80,000 was invested for 2 years at Simple Interest; what rate per annum would give the same yield?

13.

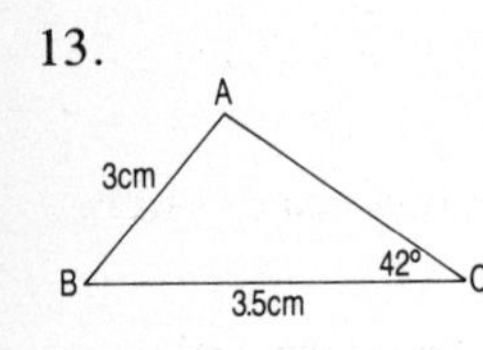

(a) Calculate the size of angle B.

(b) Calculate the length of AC.

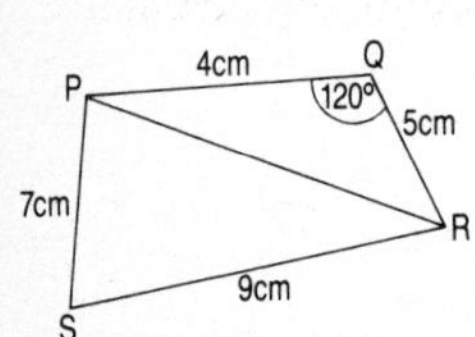

In the quadrilateral PQRS PQ = 4cm, QR = 5cm, LQ = 120°, PS = 7cm, SR = 9cm. Calculate

(c) PR

(d) angle S.

14. The marks gained in an 'A' level examination were as follows:

MARKS	1–20	21–40	41–60	61–80	81–100
FREQ.	10	40	120	90	40

(a) Make a cumulative frequency for the results.

(b) Draw the cumulative frequency graph, and estimate the median.

(c) Estimate the interquartile range.

(d) If 12% of the candidates failed, state the pass mark.

15. (a) Describe, using Sets language, the shaded areas:

(i)

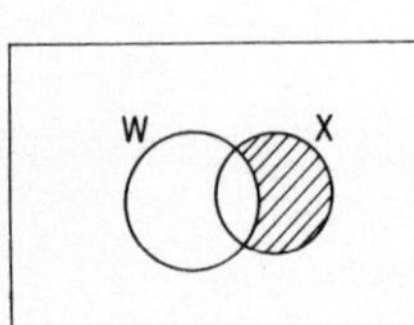

(ii)

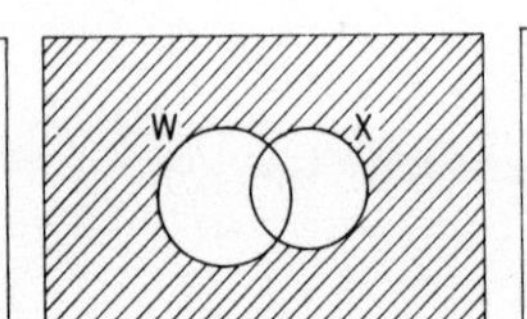

(iii)

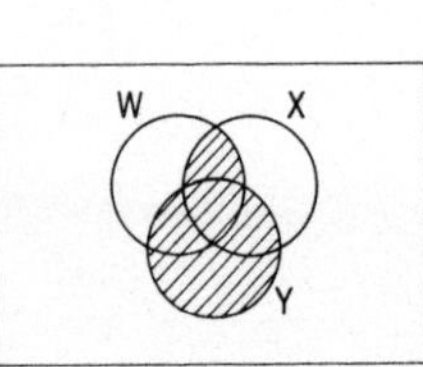

700 people were asked if they listened to LBC, Capital or Radio One – all of them listened to at least one station.

500 listened to only one station
100 listened to both Capital and LBC
70 listened to both Capital and Radio One
80 listened to both LBC and Radio One.

(700)

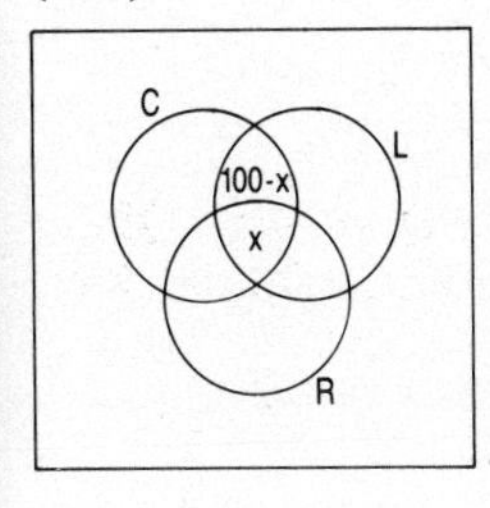

(b) By filling in two more sections of the Venn diagram calculate x.

(c) If 345 listened to Capital how many listened to only Capital?

(d) If 200 listened to only Radio One, calculate how many listened to LBC

PART II

ANSWERS TO PRACTICE PAPERS CONTAINED IN PART I

ALGEBRA AND GRAPHS

1. (a) $x = 3$ (b) $x = 1$ (c) $x = 2 \quad y = -1$
   (d) $x = 2$ or $-\frac{1}{4}$

2. (a) $-\frac{4}{5}$ (b) $2x + y + 1 = 0$
   (c) (d) $x = 1$

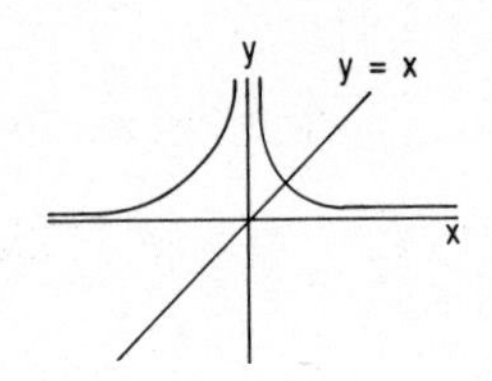

3. (a) 5, −1, −7, −7 (b) Graph (c) Gradient is 1
   (d) $x = 1\frac{1}{2}$, $(1\frac{1}{2}, -7\frac{1}{4})$

4. (a) 4 (b) (i) $4x(x + 2y)$ (ii) $(3x + 1)(x - 2)$
   (c) $x = \sqrt{\dfrac{v - n}{m}}$ (d) 10, −10

5. (a) $\pm \frac{1}{2}$ (b) $42 - 21x$ (c) $(a - 3)(x - y)$
   (d) $9 - 12x + 4x^2$; 25

6. (a) 16 (b) $x(x - 6) = 16$; 8 and 2 or − 2 and − 8
   (c) −5 (d) $x^2 - 5x + 4$ for example.

AREA AND VOLUME

1. (a) $100cm^3$ (b) $145cm^2$ (c) 1 : 4 (d) 640

2. (a) $416m^2$ (b) 4600 (c) $12.6m^2$ (d) 2,500,000

3. (a) $603cm^3$ (b) 16 : 9 (c) 3 (d) 4

4. (a) $90cm^2$ (b) 1 to 9; $10cm^2$
   (c) A solid with uniform cross section (d) 15cm

5. (a) $33.5cm^3$ (b) $50cm^2$ (c) $96cm^2$ (d) 52.3%

6. (a) $1850cm^3$ (b) $528cm^2$ (c) $2352cm^3$
   (d) r = 2.56cm

CIRCLE THEOREMS

1. (a) 100°

   (b) One reason is B = 40° + 50° = 90°: hence AOC is a diameter.

   (c) 80° (ii) 120°

2. (a) Use angles in same segment. (b) 4½

   (c) 108° (d) 18°

3. (a) 120° (b) Opposite angles add up to 180°; centre of OT.

   (c) Proof (d) $\sqrt{24}$

4. (a) 13 (b) $\sqrt{50}$ (c) 4π (d) 72°

5. (a) 80° (b) 20° (c) M 50°, L 60°, N 70°

   (d) No; a rectangle is cyclic.

6. (a) 36° (b) 18° (c) 45° (d) 45°, 45°, 90°

CONSTRUCTIONS (including bearings and scales)

1. (a)/(b) as in diagram.

   [As the △ is right angled the centre of the circle is in the middle of the hypoteneuse.]

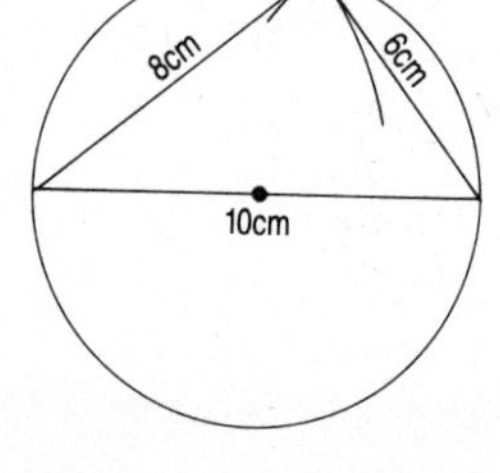

1. (c) 25 (d) 4

2. (a)/(b)/(c) as in diagram.

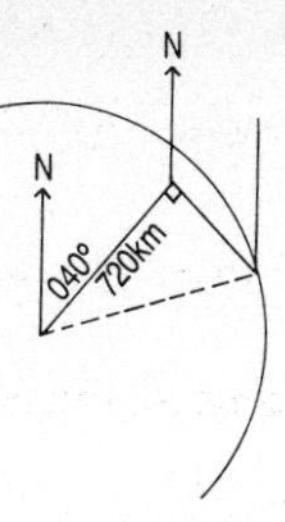

(d) 251°

3. (a) Various answers. (b) 73.7°

(c) Sketch

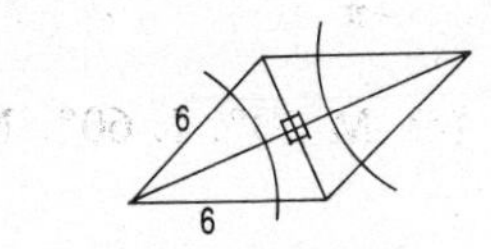

(d) $50m^2$

4.

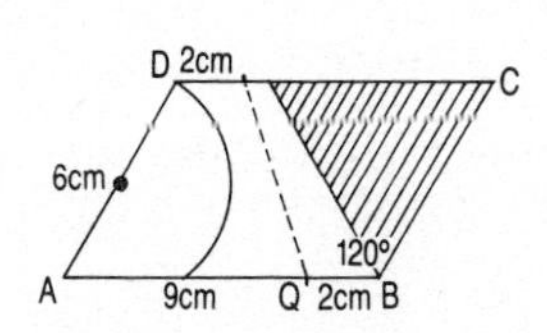

5.

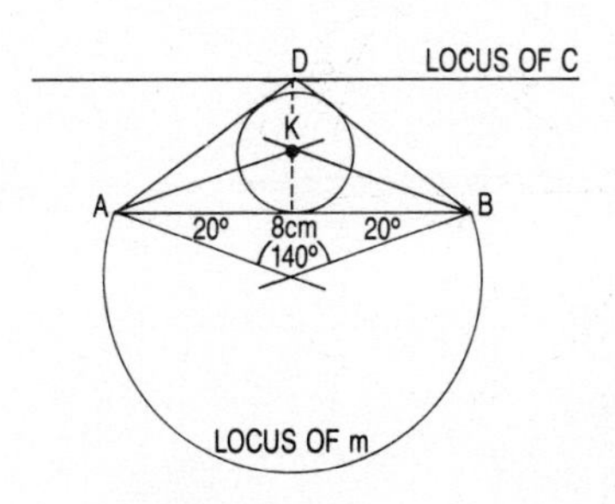

GEOMETRY OF RECTILINEAR FIGURES

1. (a) $30°$ (b) $50°$ (c) $x = 40$ (d) 20

2. (a) $95°$ (b) 1 (c) 75 (d) $30°$, 45, 105

3. (a) $2x + 2y = 180 \therefore x + y = 90$; $180 - 90 = 90$
   (b) 55 (c) $130°$ (d) $120°$

4. (a) $50°$ is not a factor of $360°$. (b) $150°$
   (c) Every square is a rhombus.
   (d) Either into 2 trapezia or 2 pentagons.

5. (a) 115 (b) $90 + \frac{k}{2}$ (c) 16 (d) 2520

6. (a) 108 (b) proof (c) $40°, 70°, 70°$ (d) $1440°$

MATRICES & TRANSFORMATIONS

1. (a)/(b) shapes; 1 : 9 (c) $\begin{pmatrix} 3 & 0 \\ 0 & 3 \end{pmatrix}$ (d) $\begin{pmatrix} \frac{1}{3} & 0 \\ 0 & \frac{1}{3} \end{pmatrix}$

2. (a) $\begin{pmatrix} -1 & 1 \\ 1 & -1 \end{pmatrix}$ (b) 180° Rotation about (0, 0)

   (c) Reflection in $y = -x$; Self inverse.

   (d) $\begin{pmatrix} 0 & 1 \\ 1 & 0 \end{pmatrix}$; Reflection in $y = x$

3. (a) $a = 2$ (b) $\begin{pmatrix} 10 & 12 \\ 18 & 22 \end{pmatrix}$ (c) $\begin{pmatrix} 0 & 1 \\ 1 & 0 \end{pmatrix}$

   (d) $x = \pm 3$; the co-ordinates are in a straight line, hence "the shape" has no area.

4. (a) (−1, 4) (1, 4), (−1, 6) (b) $m = 2$, $n = 0$

   (c) $\begin{pmatrix} -2d & -10 \\ 3d & 15 \end{pmatrix}$

   (d) $\begin{pmatrix} 2d & 10 \\ 4d & 20 \end{pmatrix}$ has a determinant of 0.

5. (a) shapes (b) $\begin{pmatrix} -1 & 0 \\ 0 & 1 \end{pmatrix}$

   (c) Translation $\begin{pmatrix} -12 \\ 0 \end{pmatrix}$ (d) 8 ; 2

6. (a) $\begin{pmatrix} 4 & -7 \\ -1 & 2 \end{pmatrix}$ (b) $\begin{pmatrix} \frac{1}{2} & 0 \\ 0 & 2 \end{pmatrix}$

   (c) ± 3 (d) Ratio of their areas.

NUMBERS AND FINANCE

1. (a) £725 (b) £40 (c) £1750 (d) 8%

2. (a) 41 or 43 or 47 (b) 10.15 (c) 3^5

   (d) $2^3 \times 3^3$; $\sqrt[3]{216} = 2 \times 3 = 6$

3. (a) $\dfrac{1 \times 30}{3} = 10$ (b) 5×10^{-2} (c) $\frac{2}{3}$

   (d) 1.6×10^5

4. (a) £120 (b) 3 (c) 45450 to 45549 (d) $2\frac{3}{5}$

5. (a) £34.48 (b) 53.8 mph (c) $\frac{5}{7}$, 71%; $(0.84)^2$

   (d) own example

6. (a) £51 (b) 12% (c) £280 (d) 80%

PYTHAGORAS AND TRIGONOMETRY

1. (a) 35 (b) 71.1° (c) $\sqrt{108}$ (d) 35.3°

2. (a) 36.9° (b) 2.4 (c) 58.7 (d) 4.60

3. (a) 20.0 (b) 123 (c) 10.8m

4. (a) 4.20m (b) 7.78m

   (c) $61^2 = 3721$
   $60^2 + 11^2 = 3600 + 121 = 3721$

   (d) 10.8

5. (a) 48.6 (b) 122m (c) $^8/_{17}$, $^{15}/_8$ (d) 0.484

6. (a) 1203m (b) 6 mins (c) 13.3°

PROBABILITY AND STATISTICS

1. (a) same number of boys as girls
   (b) different numbers of boys and girls
   (c) 87kg (d) $66\frac{1}{3}$kg

2. (a) $^{2}/_{12}$ is wrong; she should work from $^{12}/_{52}$ to get $^{11}/_{51}$
   (b) $^{11}/_{221}$ (c) $^{5}/_{6}$ (d) $^{5}/_{6}$

3. (a) 60 – 79 (b) 60% (c) 62.2 (d) 40 – 59

4. (a) $\frac{3}{8}$
   (b) What are the chances of getting at least one head (or tail)?
   (c) 22 (d) 50mm

5. (a) 1, 4 (b) $^{3}/_{14}$ (c) 250 (d) 116°

6. (a) $^{1}/_{5}$ (b) $^{21}/_{40}$ (c) $^{3}/_{40}$ (d) 15

VECTORS

1. (a) $\begin{pmatrix}13\\-3\end{pmatrix}$ (b) $\sqrt{13}$ (c) -25 (d) $\sqrt{2600}$

2. (a) $\underline{a} + \underline{b}$, $\underline{b} - \underline{a}$ (b) $\frac{1}{4}\,\underline{a} - \frac{1}{4}\underline{b}$

   (c) Compare the vectors (d) $1 : 8$

3. (a) $\begin{pmatrix}6\\2\end{pmatrix}$ (b) $\sqrt{20}$, $\sqrt{20}$

   (c) Rhombus (d) $\sqrt{40}$; square

4. (a) $1\frac{1}{2}$ (b) $\begin{pmatrix}-\varepsilon\\4\end{pmatrix}$ (c) 13

   (d) various e.g. $\begin{pmatrix}-5\\-12\end{pmatrix}$ $\begin{pmatrix}0\\13\end{pmatrix}$ $\begin{pmatrix}-13\\0\end{pmatrix}$ etc.

5. (a) $\frac{1}{3}\,\underline{a}$, $-\frac{2}{3}\,\underline{a}$ (b) $\underline{b} - \underline{a}$ (c) $\frac{1}{3}(\underline{b} - \underline{a})$

   (d) $8 : 9$

6. (i) (a) For example $|\overrightarrow{AC}| =$
   $\sqrt{98}$ and $|\overrightarrow{AB}| + |\overrightarrow{BC}| = \sqrt{32} + \sqrt{18} = \sqrt{98}$

   (b) $\begin{pmatrix}1\\0\end{pmatrix}$ (c) $a = b = 0$

   (d) $a = kb$, where k is a constant.

MIXED EXAMPLES 1

1. (a) 1.369×10^9 (b) 8×10^{-3}

2. (a) 15% (b) £6.88

3. (a) $-\frac{1}{2}$ (b) (2, 2)

4. 3, $-\frac{2}{3}$

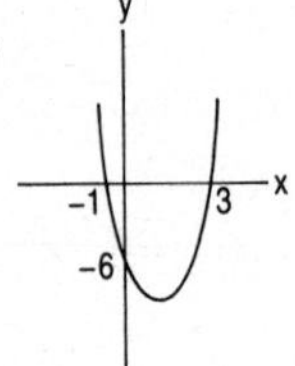

5. (a) $2 \times 3 \times 5 \times 7$ (b) 16, 96 (c) $\frac{3}{16}$

6. (a) $\begin{pmatrix} -1 & 0 \\ 0 & 1 \end{pmatrix}$ (b) Reflection is y-axis

7. 4

8. $11\frac{1}{9}$

9. 8 : 27

10. various e.g.

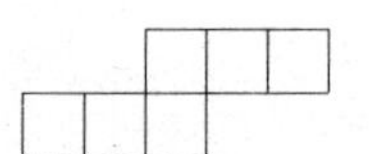

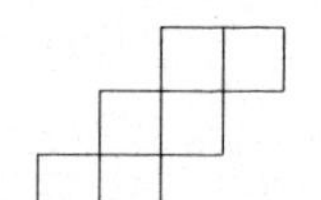

11. (a) 9.77 (b) 39.1

12. 6 or -8

13. 48

14. -1, $-5\frac{1}{2}$

15. (a) 73.7° (b) 84cm^2

16. (i) $\underline{b} - \underline{a}$ (ii) $\frac{4}{5}\underline{a} - \underline{b}$

17. (a) 0.7 (b) 14km^2

18. 190°

19. (i) $\begin{pmatrix} 5 \\ 1 \end{pmatrix}$ (ii) 45°

20. (a) 160° (b) 50°

21. (i) $x < -2\frac{1}{2}$, (ii) $x < -25$

(iii) −2, −1, 0, 1 12

22. 19

23. (i), (iii), (v)

24. Sketch

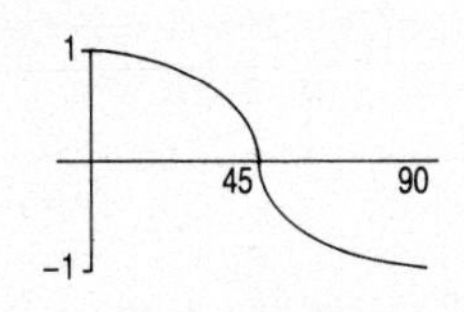

25. ¼

26. (a) −35 (b) ± ⅓

27. 150°

28. $y = \frac{1}{4}x - 2$

29. $\sqrt{320}$ [N.B. 18 is **WRONG**]

30. £5832

MIXED EXAMPLES 2

1. (a) Graph (b) $8\frac{1}{4}$ (c) Gradient − 5
   (d) $c = 12$, $b = 1$, $a = -1$

2. (a) (i) 268cm^3 (b) $2144/5\text{cm}^3$ (ii) 2cm

3. Radius = 6.7

4. (a) $(-4, 4)$, $(-2, 5)$, $(0, 2)$
   (b) $(0, -2)$, $(2, 1)$, $(4, 0)$, $\begin{pmatrix} 1 & 0 \\ 0 & -1 \end{pmatrix}$
   (c) $\begin{pmatrix} -4 \\ -3 \end{pmatrix}$ (d) Reflection in $y + x = 0$

5. (a) 3^6 (b) 32 (c) $1.2 \times 10'$ (d) 50000, 49680

6. (a) 290° (b) $0 = 360 - 2A$ (c) 160° (d) 40°

7. (a) 50° (b) 9 (c) 20 (d) false

8. (a) $\begin{pmatrix} 1 \\ -11 \end{pmatrix}$ $\sqrt{34}$ (b) (4, 5) (c) $\begin{pmatrix} 3 \\ 3\frac{1}{2} \end{pmatrix}$
   (d) Adjacent sides are $\sqrt{8}$ and $\sqrt{41}$

9. (a) 135° (b) $^4/_9$ (c) 8.3 years (d) $^{49}/_{87}$

10. (a) 3.42, 9.40 (b) 36.1° (c) $^{12}/_{37}$ (d) 9.14

11. (a) 8, 66 (b) $-\frac{2}{3}$ (c) $-3x^2 - 4$ (d) 0, −3

12. £95400; 9.625%

13. (a) 86.7° (b) 4.48 (c) $\sqrt{61}$ (d) 56.8

14. (a) 10, 50, 170, 260, 300

 (b) Graph; 58

 (c) Approx. 72 − 46 = 26

 (d) Approx. 34

15. (a) (i) $W' \cap X$ (ii) $(W \cup X)'$ (iii) $Y \cup (W \cap X)$

 (b) x = 25

 (c) 200

 (d) 255

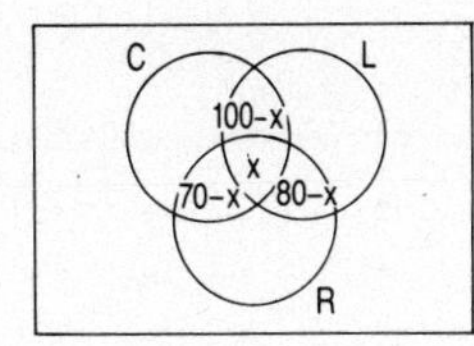

PART III

GCSE PRACTICE PAPERS

TIME ALLOWED
1 HOUR EACH PAPER

INTRODUCTION AND CONTENTS

There are ten papers, each designed to take an hour and testing a different topic (or a small group of topics). *Supplementary* questions follow individual papers for two reasons:-

(a) for use by students who finish the paper early

(b) to provide substitute questions should it be decided that one (or more) questions in a paper should be replaced for test purposes.

There are also two longer papers of mixed examples for extra practice before Mock Examinations or the main examination. They also provide additional material should you wish to "create" further practice papers.

Paper

ALGEBRA AND GRAPHS

TIME 1 HOUR

1. (a) Solve the equation: $\frac{x}{3} - 1 = 5$

(b) The perimeter of the rectangle below is 20cm; calculate x

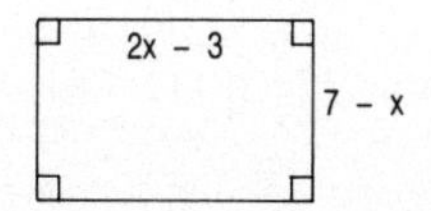

(c) Solve the simultaneous equations:
$3x - 2y = 5$ and $5x + 3y = 2$.

(d) Solve the quadratic equation: $6x^2 - 13x + 6 = 0$.

2. (a) The equation of a straight line is $3x - 2y = 4$; what is its gradient?

(b) What is the equation of the straight line that passes through the points (3, 2) and (0, −1)?

(c) Draw a neat sketch of the graph $y = \frac{1}{x}$,

for $x = \pm 4, \pm 3, \pm 2, \pm 1, \pm \frac{1}{2}$

(d) What is the equation of the line of symmetry that intersects the graph twice? Write down the co-ordinates of the points of intersection.

3. (a) Copy and complete the table for the graph of $y = 2x^2 - 6x$

x	−1	0	1	2	3	4
y				−4		8

(b) On a sheet of graph paper choose suitable axes and draw the graph.

(c) Draw a tangent to the curve at $x = 2\frac{1}{2}$; hence find the gradient at this point.

(d) Use the graph to solve the equation: $2x^2 - 6x + 1 = 0$

4. (a) $t = 3p^2$ Find t when $p = -1$

(b) Rearrange the formula in (a) to make p the subject.

(c) $\dfrac{F - 32}{9} = \dfrac{C}{5}$

Calculate F when $C = 10$

(d) By writing $C = F$ in the equation in (c), calculate the value for which $C = F$.

5. (a) x varies as the square of y, and x is 4 when y is 48. Find x when y is 192

(b) Simplify: $3(2x - 3) - 5(2 - 3x)$

(c) Factorise the expression $3x^2 + 6xy$

(d) Make x the subject: $3x + m = n - 4x$

6. (a) There are 7 more boys than girls in a class; if there are x girls in the class and 29 pupils altogether, form an equation in x and solve it.

(b) The length of a rectangle is 7 units more than its width; let the width be x units, and write down an expression for the area of the rectangle. Given that the area is 30 square units, calculate the value of x.

(c) $y = 2x^{1/3}$; find y when $x = 8$

(d)

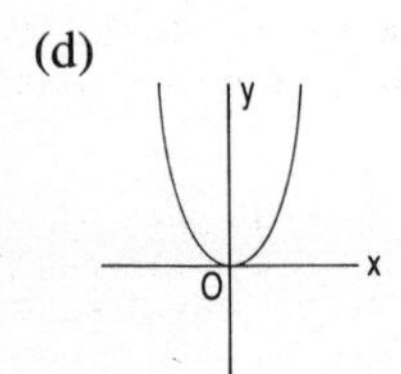

The sketch-graph represents the relationship $y = ax^2$, (where a is a constant). The point $(-3\frac{1}{2}, 1\frac{3}{4})$ is on the graph; calculate a giving your answer as a fraction in lowest form.

ALGEBRA AND GRAPHS
SUPPLEMENTARY QUESTIONS

1. Factorise $16x^2 - 25y^2$; if the sum of the two factors is 20 find x.

2. Rearrange the formula $x = \dfrac{ap + q}{bp + q}$ to make p the subject.

3. Make a sketch of the graph of $y = x^2 - 3x - 4$, showing clearly where it intersects the axes. Estimate the area between the graph and the x-axis, by using trapezia and triangles.

4. y varies as x^2, and y is $^1/_{36}$ when x is $^1/_9$; Find x when y is 4.

AREA AND VOLUME
TIME 1 HOUR

1.

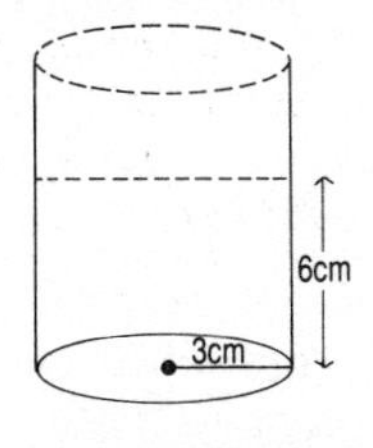

(a) A cylindrical glass has a base radius 3cm. It is filled with juice to a height of 6cm. Calculate the volume of the juice.

(b) Two sugar cubes, each with a 2cm edge, are dropped into the juice. Calculate in mm by how much the juice will rise.

(c) In another cylinder the diameter of the base is equal to twice its height; let the diameter be 2r and write down an expression for the volume of this cylinder in terms of r.

(d) Given that the volume of this cylinder is $64\pi cm^3$, calculate r.

2. The radius of a sphere is 3cm

   (a) Calculate the volume of the sphere, giving your answer to 1 decimal place.

   (b) A second sphere has a radius of 4.5cm. Express the ratio of the volume of the smaller sphere to the volume of the larger sphere in the form a : b where a and b are whole numbers.

   (c) What is the surface area of the sphere of radius 3cm?

   (d) Express the ratio of the surface area of the smaller sphere to the surface area of the larger sphere in the form p : q where p and q are whole numbers.

3. (a) A cuboid has a square base of side x and a height of 10cm. Write down, in terms of x, an expression for the total surface area of the cuboid.

   (b) Given that the volume of the cuboid is $640cm^3$, calculate the value of x.

   The cross-section of a prism is a trapezium as shown below

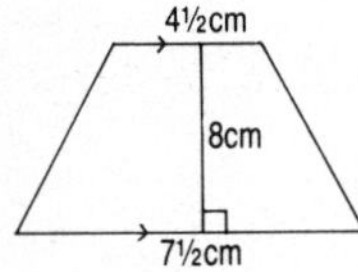
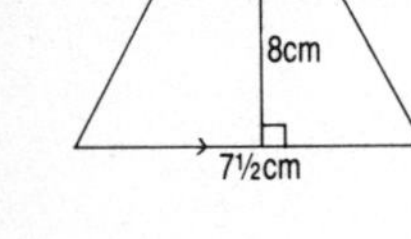

   (c) Calculate the area of the cross-section.

   (d) A similar prism has height 16cm and volume $576cm^3$. Calculate the length of the first prism.

4.

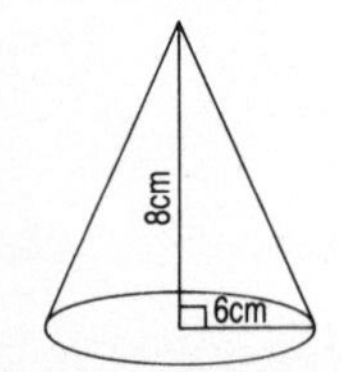

   (a) Calculate the volume of the cone, giving your answer to 3 significant figures.

   (b) A small cone of height 2cm is sliced off the top of the given cone. Calculate the volume of the remaining frustum.

   (c) Calculate, in terms of π, the curved surface area of the given cone.

   (d) State in simplest terms the ratio of the curved surface area of the smaller cone to that of the original.

5. (a) The area of a circle is $120cm^2$; calculate its radius.

   (b) 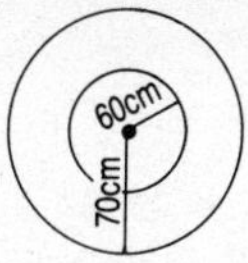 Calculate the area of the ring between two concentric circles of radii 60cm and 70cm.

   (c) The area of a sector of a circle of radius 8cm is $100cm^2$. Calculate the sector angle.

   (d) Calculate the area of the remaining part of the circle from which the sector was cut (in (c)).

6. 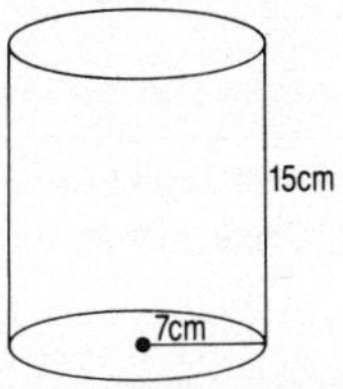 (a) Calculate the curved surface area of the cylinder.

   (b) If a litre of water is poured into the cylinder, what height will it reach.

ABCD is a rhombus with diagonals 9cm and 7cm.

   (c) Calculate its area, stating any assumptions you make about its diagonals.

   (d) The area of a similar rhombus is $8064cm^2$. Find the length of its diagonals.

AREA AND VOLUME
SUPPLEMENTARY QUESTIONS

1. 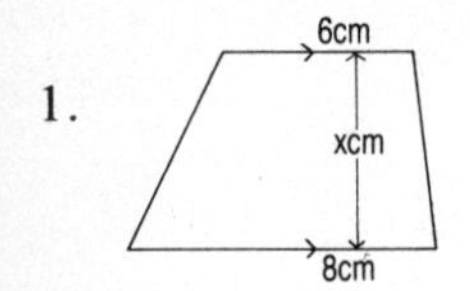

The area of the trapezium is $16cm^2$; Calculate the value of x to 2 decimal places.

2. 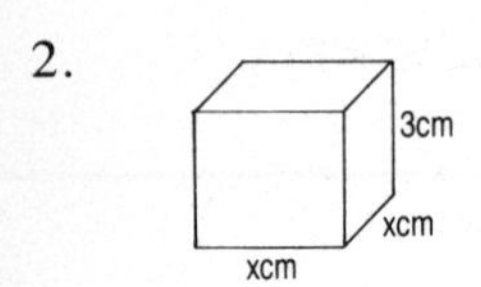

The volume of this square-based cuboid is $243cm^3$; Calculate x and find the total surface area of the box.

3. A sphere and a cone have the same volume; if the base radius of the cone is the same as the radius of the sphere, express the height of the cone in terms of its radius.

4. Using h for the height, and r for the radius, write down an expression for the area of the curved surface of a cylinder. Given that $h = 2r$ and the area is $300cm^2$, calculate r.

CIRCLE GEOMETRY (including intersecting chords and the length of an arc)

TIME 1 HOUR

(Note: In this paper the point marked O. is the centre of the circle)

1.

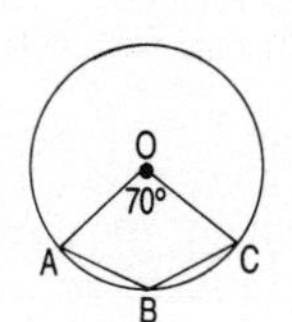

(a) Calculate the size of angle ABC.

(b) Extend AO to meet the circumference at D; what is the size of angle OCD?

(c) If also BA = BC, calculate the size of angle OCB.

(d) What is the size of angle BDC?

2. In the diagram the diameter AX is produced to D. DC meets BY produced at C, and is parallel to BA. The angle BYX is 20° and angle C is 60°.

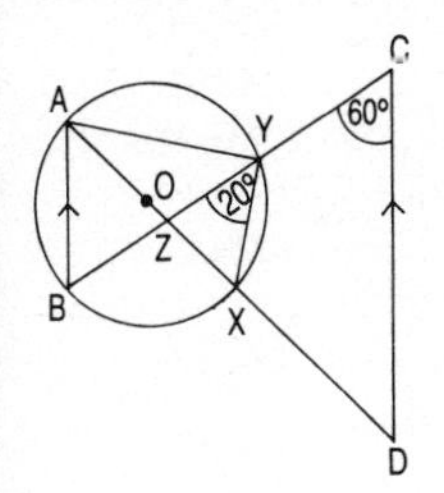

(a) Calculate the size of angle AZB.

(b) Calculate the size of angle YAX.

(c) Explain why the points A, B, C, and D cannot be the vertices of a cyclic quadrilateral.

(d) Write down three similar triangles.

3.

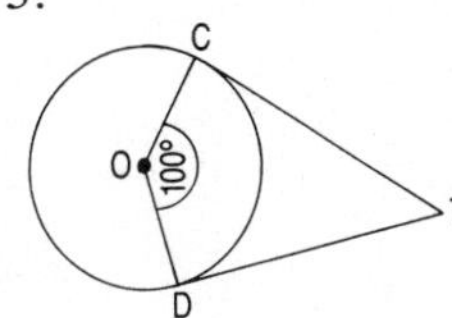

(i) In the diagram TC and TD are tangents.

(a) Calculate the size of angle T.

(b) Join TO and extend it to meet the circumference at E. Calculate the size of each angle of triangle CET.

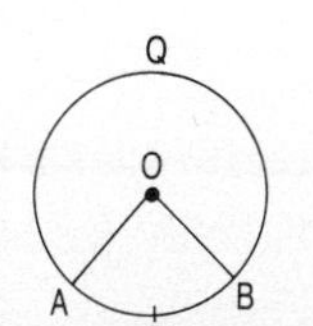

(ii) The angle AOB is 100°; P is a point on the minor arc AB and Q on the major arc.

(c) Express in simplest form the ratio of the length of the arc APB to the length of the arc AQB.

(d) Given that the arc APB is 12cm, calculate the radius of the circle.

4.

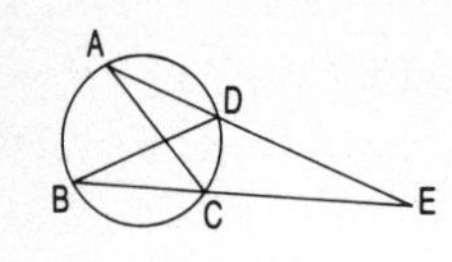

(i) In the diagram the chords AD and BC are produced to meet at E.

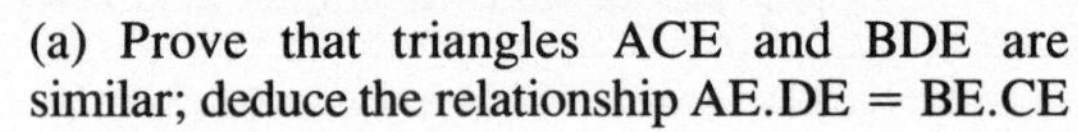

(a) Prove that triangles ACE and BDE are similar; deduce the relationship AE.DE = BE.CE

(b) If in the diagram angle E is 20° and angle A is 40°, calculate the size of angle ADB.

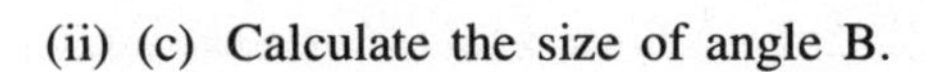

(ii) (c) Calculate the size of angle B.

(d) Prove that a trapezium which is cyclic is also isosceles.

5.

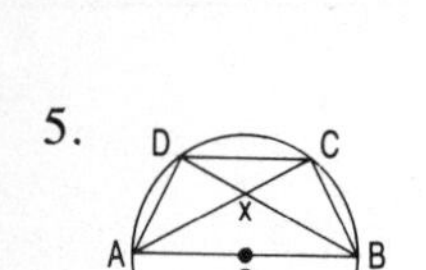

(i) In the diagram AOB is a diameter; the chords AC and BD intersect at X. Angle ADC is 130° and angle AXB is 120°.

Calculate

(a) the size of angle DCB

(b) the size of angle BOC.

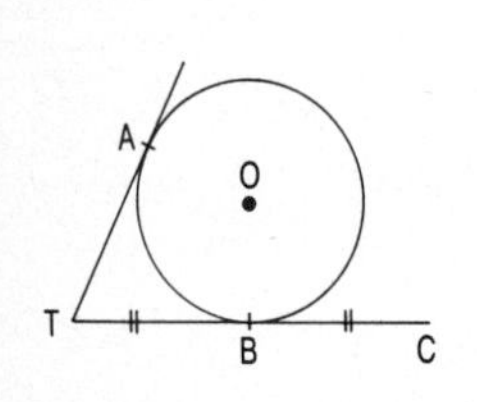

(ii) TA and TB are tangents to the circle with TB extended to C; BT = BC

Given that angle AOB = 122°, calculate the size of

(c) angle TCO

(d) angle ABT

6.

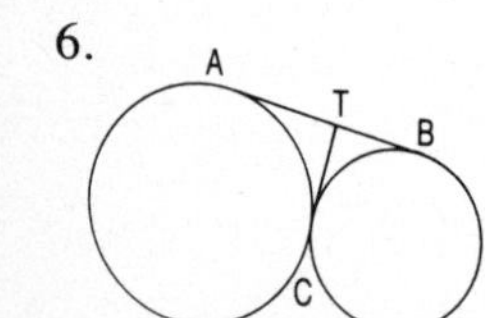

In the diagram TA, TB and TC are tangents, with A, T and B in a straight line.

Prove

(a) that TA = TB

(b) that angle ACB is 90°.

(c)

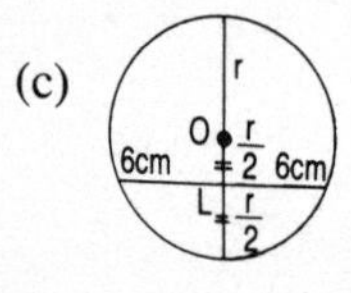

Prove that the radius, r, is $\sqrt{24}$

(d)

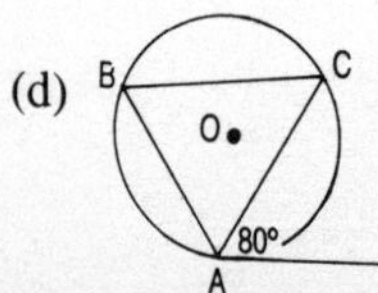

Given that TA is a tangent, calculate the size of angle OCA.

CIRCLE GEOMETRY
SUPPLEMENTARY QUESTIONS

(Note O. is the centre of the circle)

1.

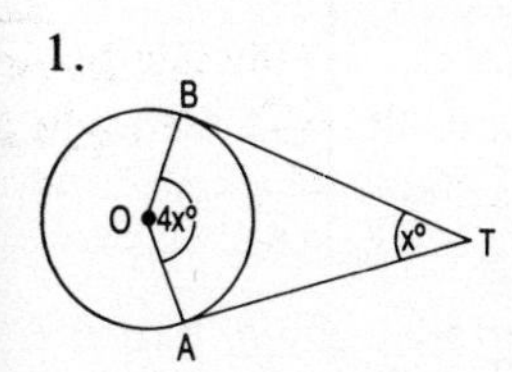

In the diagram TA and TB are tangents.

Calculate the size of x.

Explain why AOBT is cyclic; under what circumstances would the centre of the circumscribing circle lie on the arc AB?

2.

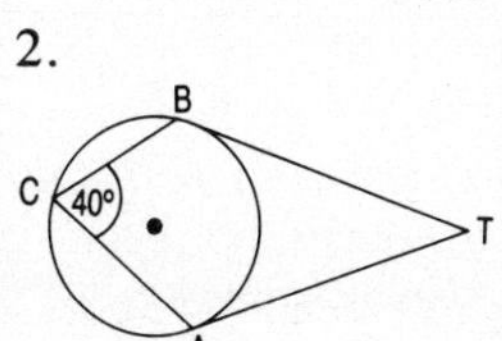

If TB and TA are tangents, and angle C is 40°, calculate the size of angle T.

3.

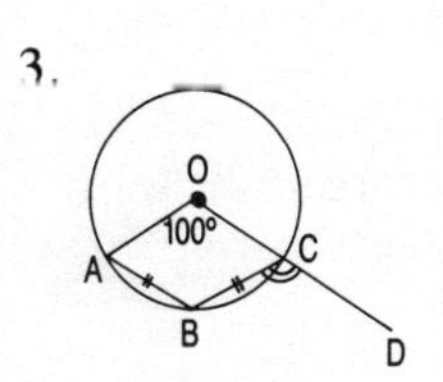

Given that AB = BC, and the angle O is 100°, calculate the size of the exterior angle BCD.

4.

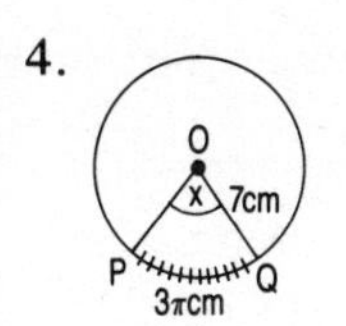

The arc PQ is 3πcm; the radius is 7cm. Calculate the size of angle x.

CONSTRUCTIONS
(including bearings and scales)
TIME 1 HOUR

1. (i) Construct a triangle ABC with AB = 8cm, BC = 6.5cm and angle B = 30°. Indicate by shading the set of points which are less than 4cm from B and less than 3cm from AC.

   (ii) A distance of 4km is represented on a map by 1.25cm. Express the scale of the map in the form 1 : n.

   What is the area in cm^2 on the map of a ranch which is $128km^2$.

2. Two ports A and B are 30km apart with B due East of A; a ship sails North from B for 45km, and then alters its course to a bearing of 215°. It continues on this bearing until it is 20km from A.

   (a) Using a scale 5cm = 1km make an accurate diagram of the ship's course.

   (b) Draw clearly the locus of all points which are 20km from A.

   (c) Mark with the letter C the final position of the ship; from A a small boat is sent out to the ship at 7km/h. On what bearing should the small boat sail?

   (d) How long, to the nearest minute, will it take to reach the ship?

3. The scale on a map is 1 : 40 000

   (a) What distance does a line 12.5mm on the map represent?

   (b) What will a river, 9.6km, measure on the map?

B is on a bearing of x° from A, make 3 sketch diagrams to represent this; x must be acute in one, obtuse in another, and reflex in the third.

Express, in each case, the back-bearing from B to A in terms of x.

4. Draw a trapezium ABCD with DC = 11cm, AD = 8cm and angle D = 70°. AB = BC, and AB is parallel to DC.

Indicate clearly how you located the point B.

Indicate by shading the set of points which are closer to D than C.

Mark with an X the point which is 2cm from DC and equidistant from A and D.

5. (i) The height of a triangle is 5cm, and its base angles are 60° and 50°. Construct the triangle accurately. (No credit will be given for answers based on trigonometrical calculations.)
What is the area of the triangle you have drawn, to 1 decimal place?

(ii) Draw a straight line AB of length 8cm; on one side construct a right angled triangle ABC with angle C equal to 90°, and the area of the triangle $16cm^2$. On the other side construct the locus of M where angle AMB is 80°.

CONSTRUCTIONS
(including bearings and scales)
SUPPLEMENTARY QUESTIONS

1. The scale on a map is given as 1cm = 8km.

   (a) Write this in the form 1 : n

   (b) What is the area on the map of a lake of area $72km^2$.

   (c) How far apart are two places which are 75mm apart on the map?

2. A ship sails 600km on a bearing 325°; it then sails NE until it is directly North of its starting point. Using a scale of 1cm to 75km make an accurate diagram of the ship's journey.

How long, to the nearest minute, will it take the ship to return to its starting point if it then travels at 30km per hour due south?

3. Draw a circle of your own; now draw an equilateral triangle to circumscribe the circle.

4. On one side of a line AB, 9cm long, construct the locus of a moving point C such that the area of the triangle CAB is $13\frac{1}{2}cm^2$. Mark on your locus two points C_1 and C_2, where angle AC_1B = angle AC_2B = 90°.

Measure the distance C_1C_2.

GEOMETRY OF RECTILINEAR FIGURES

TIME 1 HOUR

1. (a) 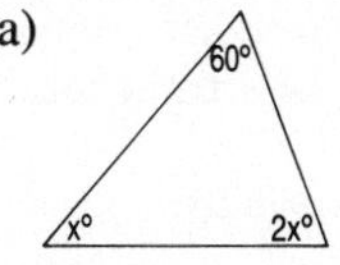

Calculate the value of x in the triangle on the left.

(b) 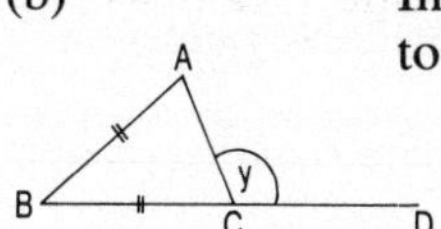

In the triangle ABC, BA = BC; BC is extended to D. Express angle B in terms of y.

(c) Each interior angle of a regular polygon is 140°; calculate how many sides the figure has.

(d) The ratio of the size of each interior angle of a regular polygon to each exterior angle is 5 : 1. What is the sum of the degrees of the polygon?

2. 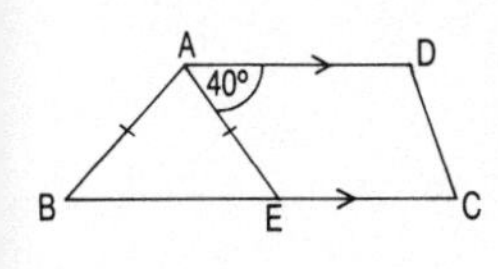

(i) Given that AD is parallel to BC and AB = AE,

(a) Calculate the size of angle BAE?

(b) If also angle D is 140°, what is the distinct name for the quadrilateral ADCE?

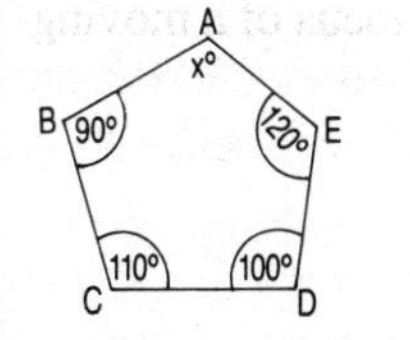

(ii) (c) Calculate the size of the angle marked x.

(d) The interior angles of another polygon add up to double that of a pentagon. How many sides has that polygon?

3. 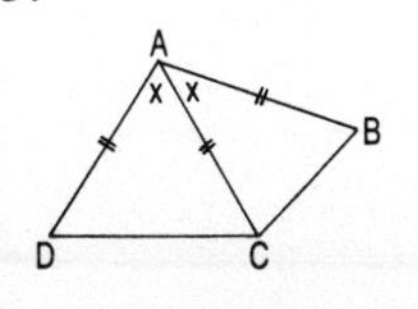

(i) In the quadrilateral ABCD AD = AB = AC, and AC bisects angle A;

(a) Prove that CD = CB.

(b) If also AB is parallel to DC, deduce the size of angle D.

(ii) (c) Name a parallelogram with two unequal lines of symmetry.

(d) Is it correct to say a square is a parallelogram?

4. ABCDEFGH is a regular octagon, with alternate points joined.

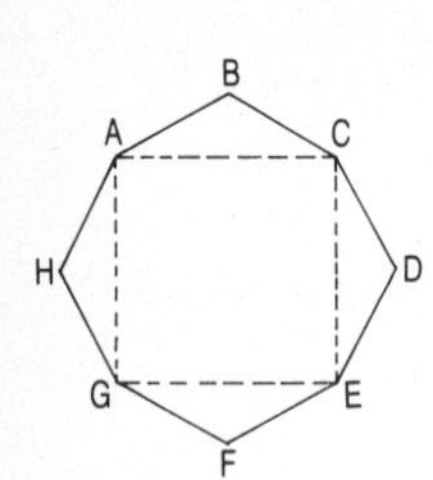

(a) Write down the size of angle B.

(b) Prove that triangles AHG and CDE are congruent.

(c) Show that angle GAC is a right angle.

(d) Prove that ACEG is a square.

5. (a) How many lines of symmetry has a regular pentagon?

The diagram shows a regular pentagon and a square standing on opposite sides of ED.

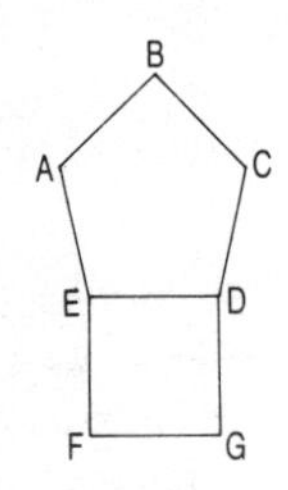

(b) Calculate the size of the obtuse angle CDG.

(c) Calculate the size of angle GEC.

(d) Prove that the diagonal AC is parallel to ED.

6. ABCDEF is a regular hexagon with diagonals intersecting at O.

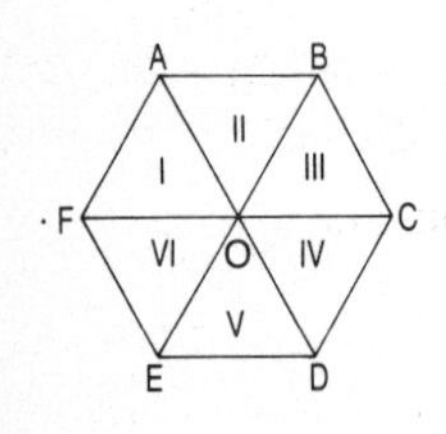

Which transformation takes

(a) △I to △III with A→B, F→C, and O unmoved?

(b) △I to △VI with A→F, F→E, and O unmoved?

(c) △IV to △VI with O→F, C→O and D→E?

(d) △III to △II where neither B nor O is mapped on to itself?

GEOMETRY OF RECTILINEAR FIGURES
SUPPLEMENTARY QUESTIONS

1. The sum of the degrees of a polygon is 3600°; calculate how many sides it has.

2. 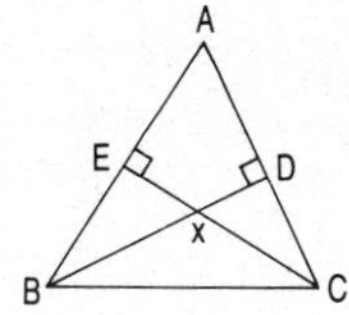

ABC is an isosceles triangle with AB = AC; BD and CE are perpendicular to AC and AB respectively. Prove that triangles AEC and ADB are congruent.

Express angle BXC in terms of angle A.

3. 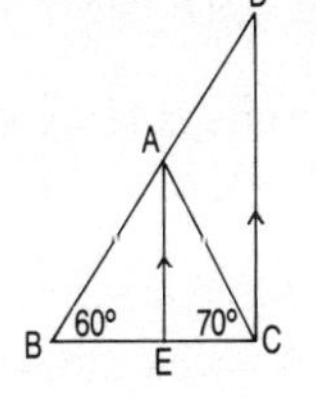

In the triangle ABC, angle B is 60°, and angle C is 70°; AE is the bisector of angle A. D is a point on BA produced such that DC is parallel to AE. Show that triangle ADC is isosceles.

4. 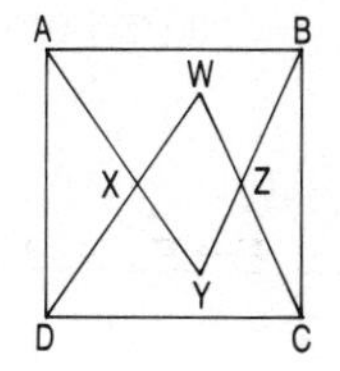

Equilateral triangles are drawn on the sides DC and AB of the square ABCD.

Prove that WXYZ is a rhombus.

MATRICES AND TRANSFORMATIONS

TIME 1 HOUR

1.

$$A = \begin{pmatrix} 0 & -3 \\ 1 & 1 \end{pmatrix} \qquad B = \begin{pmatrix} 1 & 3 \\ 0 & -1 \end{pmatrix}$$

(a) Calculate A − B

(b) Find the inverse of A, and the inverse of B.

(c) Calculate $(A + B)^2$

(d) If A − 3C = B, calculate the matrix C.

2. (i) A = (1, 3), B = (−1, 1), C = (−3, −3)

(a) On graph paper draw triangle ABC; draw also its enlargement with scale factor ½, centre (0, 0).

(b) Apply the matrix $\begin{pmatrix} 0 & -1 \\ -1 & 0 \end{pmatrix}$ to the original triangle. Plot the three new co-ordinates A′, B′, and C′. Describe fully the transformation which has taken place.

(c) Reflect triangle ABC in the x-axis, and give the matrix of this transformation.

(ii) What transformation maps (3, 2) onto (2, 3) and (−1, −4) onto (−4, −1)?

3.

	TEST 1	TEST 2
RANU	80	70
PAUL	50	60

The matrix shows the results of two pupils in two tests, during First Term.

(a) Copy and complete the matrix multiplication below, which gives the average mark for the two tests for each pupil.

$$\begin{pmatrix} 80 & 70 \\ 50 & 60 \end{pmatrix} \begin{pmatrix} \quad \end{pmatrix} = \begin{pmatrix} \quad \end{pmatrix}$$

(b) During Second Term the teacher set two more tests; Ranu's marks were unchanged, but Paul's improved by 20%. Copy and complete the matrix multiplication to show this:

$$\begin{pmatrix} \quad \end{pmatrix}\begin{pmatrix} 80 & 70 \\ 50 & 60 \end{pmatrix} = \begin{pmatrix} \quad \end{pmatrix}$$

(c) Evaluate: $\begin{pmatrix} 3 & 2 \\ 4 & 6 \\ 1 & 3 \end{pmatrix}\begin{pmatrix} -1 \\ 1 \end{pmatrix}$

(d) Simplify: $\begin{pmatrix} \frac{1}{2} & \frac{1}{2} \\ \frac{1}{2} & \frac{1}{2} \end{pmatrix}^2$

4.

$M = \begin{pmatrix} -1 & 0 \\ 0 & 1 \end{pmatrix}$ $\qquad N = \begin{pmatrix} 0 & -1 \\ 1 & 0 \end{pmatrix}$

(a) Calculate MN

(b) What transformation does the matrix M represent?

(c) What transformation does the matrix MN represent?

(d) Calculate $(MN)^{-1}$

5. (i) Given $K = \begin{pmatrix} 0.8 & -0.6 \\ 0.6 & 0.8 \end{pmatrix}$

(a) Calculate the determinant of K.

(b) By applying the matrix K to the square with co-ordinates (0, 0), (10, 0), (10, 10), and (0, 10), or otherwise describe fully the transformation it represents.

(ii) Draw two triangles

ABC : (7, 6), (3, 2), (5, 5)
DEF : (−5, 6), (−1, 2), (−4, 4)

How can △ABC be mapped on to △DEF?
{Use 2 transformations.}

6. (i) $P = \begin{pmatrix} 2 \\ -1 \end{pmatrix}$, $Q = \begin{pmatrix} -3 \\ 2 \end{pmatrix}$

(a) If A = (−2, 3), state its position after the translation P.

(b) $3P + nQ = \begin{pmatrix} 6 \\ x \end{pmatrix}$. Calculate n and x.

(ii) Given that M = (3, d)

(c) Calculate PM

(d) Show that (P + Q)M is singular, i.e. that its determinant is zero.

MATRICES AND TRANSFORMATIONS SUPPLEMENTARY QUESTIONS

1. Using the unit square, or a shape of your own describe fully which two transformations the matrix

$$\begin{pmatrix} 2 & 2 \\ -2 & 2 \end{pmatrix}$$ represent.

2. Find a and b where $\begin{pmatrix} 2 & 5 \\ 1 & 3 \end{pmatrix}\begin{pmatrix} a \\ b \end{pmatrix} = \begin{pmatrix} 14 \\ 8 \end{pmatrix}$

3. What transformation does the matrix $\begin{pmatrix} -1 & 0 \\ 0 & -1 \end{pmatrix}$ represent?

4. Which transformation maps (1, 2) onto (−2, −1) and (−3, 4) onto (−4, 3)?

NUMBERS AND FINANCE

TIME 1 HOUR

1. (i) (a) Calculate 12% of £87, giving your answer to the nearest 10p.

   (b) A bill including VAT at 15% came to £9.20; how much was the bill before VAT was included?

   (ii) A stallholder buys 30 pineapples every day at 70p each; she hopes to make 20% profit on her outlay;

   (c) What should she ask for each pineapple?

   (d) On a particular day she sold ⅔ of her stock at the desired price, and the remainder at such a price that she made an overall profit of 10%. What price did she average for each pineapple in the remaining ⅓ of her stock?

2. (a) Write down two prime numbers between 30 and 40.

   (b) Simplify: $\dfrac{2^5 \times 3^4}{2 \times 3^2}$ giving your answer in the form a^2.

   (c) Express 310 as the product of primes.

   (d) Write 1600 as the product of prime powers.

3. (a) Give an estimate of the following: $\dfrac{(3.1)^2 \times (1.95)^3}{\sqrt{143}}$

   Show your method; no credit will be given for calculator answers.

   (b) Express 8^3 as a power of 2.

   (c) Express 3.2 x 1500 in standard form.

   (d) Express $\sqrt{\dfrac{1.96}{10^{16}}}$ in standard form.

4. (a) Jane invested £750 for 4 years at 7½% per annum, simple interest. What did her investment amount to at the end of the period?

   (b) Hasmita invested a sum of money for 5 years at 7½%; at the end of the period the simple interest gained was £750. How much did she invest?

   (c) A tourist converted £250 into Greek drachmas and received 60000 drs. What was the rate of exchange on that day? By the end of her holiday the pound had 'strengthened' by 5% against the drachma; how much sterling will she get back for 3780 drs?

   (d) What is meant by saying 'inflation is falling'?

5. In a school, lessons are taught in 35 minute modules.

   (a) On Mondays PE has 3 modules in a row and finishes at 3.30 p.m. At what time does PE start?

   (b) Mathematics is allocated 5 periods a week in the 4th year and 6 in the 5th year; taking the 4th year to be 38 weeks long and the 5th to be 30 weeks, calculate the average number of minutes a week spent on Mathematics over the 4th and 5th years.

   (c) Simplify: $1\frac{2}{3} \times 4\frac{1}{5}$

   (d) The following argument is false:

   "if you increase the sides of a square by 10%, the areas will also be increased by 10%".

   Choose a square of your own to find the correct increase in the area.

6. (a) Taking 1 inch = 2.54cm, calculate to 1 significant figure the number of inches in a metre.

   (b) The attendance at a concert was given as "50,000 to the nearest thousand"; state the possible limits of the crowd.

   (c) A sum of money was divided in the ratio 7 : 8 : 10. The difference between the largest and the smallest shares was £540. What was the sum of money?

   (d) Divide 1.2 by 60, giving your answer as a fraction.

NUMBERS AND FINANCE
SUPPLEMENTARY QUESTIONS

1. Calculate the Compound Interest on £700 for 3 years at 8%. Give your answer to the nearest pound. Find what rate of Simple Interest would yield the same amount over the same period of time.

2. A company's profits rose from £80 000 in 1987 to £92 000 in 1988; Calculate the increase as a percentage. If the 1987 profits were 20% more than those of 1986, find how much profit the company made in 1986.

3. $175 \times 10^8 = 7/y$. Find y in standard form.

4. If each side of a square is reduced by 20%, what is the reduction in the area as a percentage?

PYTHAGORAS AND TRIGONOMETRY

TIME 1 HOUR

1\.

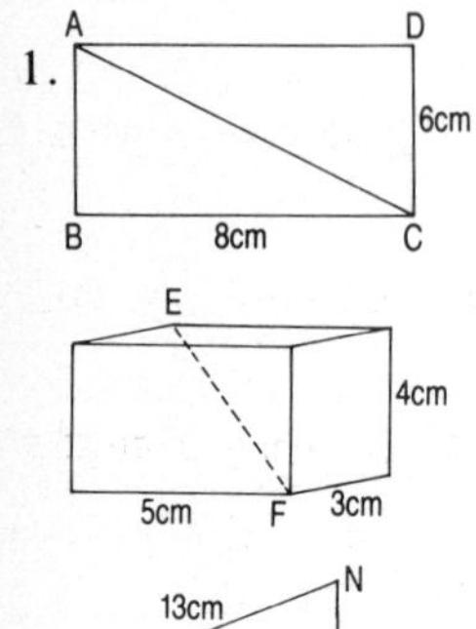

(i) (a) Calculate the length of the diagonal AC, in the rectangle ABCD.

(b) Calculate the size of the acute angle between the diagonals.

(ii) (c) The diagram is of a cuboid with dimensions shown; calculate the length of EF, giving your answer to the nearest cm.

(d) Calculate the area of the isosceles triangle LMN.

2\. In the quadrilateral ABCD $\hat{C} = 90°$; $A\hat{D}B = 90°$.

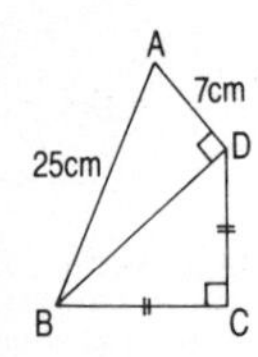

(a) Find the length of BD.

(b) If BC = CD, calculate the length of BC.

(c) Calculate the size of angle A.

(d) Calculate the size of angle ABC.

3\. A vertical flagpole AB is 18m high; it is secured to the ground by two ropes AE and AW each 25m long. The points W, B, and E are in a straight line.

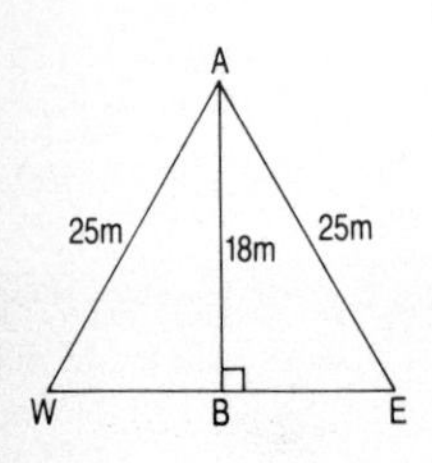

(a) Calculate the distance WE.

(b) What is the angle of elevation of the top of the pole from E?

F is a point ⅔ of the way up the pole, from which 2 new ropes FW and FE are fastened.

(c) Calculate the length of FE.

(d) Calculate the size of angle AFE.

4. A child sitting at the top of a cliff, 600m high, can see a boat 9km due East from the base of the cliff.

(a) Express in simplest form the ratio 600m : 9km.

(b) Calculate the angle of depression of the boat from the child's position.

The boat now sails due North for 15 minutes at a speed of 15km/h, while the child stays in the same position.

(c) How far is the boat now from the point at the base of the cliff directly below the child?

(d) What is the angle of elevation of the child's position from the boat now?

5. (a) 3 Sin x = 1.44. Calculate x

(b) Calculate the length of the side marked y.

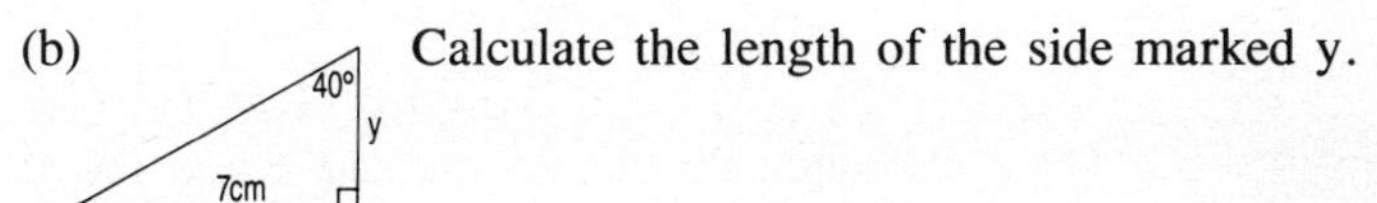

Two young trees of heights 150cm and 180cm were planted at a distance 80cm from each other.

(c) What was the angle of elevation of the top of the taller tree from the top of the smaller tree?

(d) During a storm the smaller tree was blown over until it came to rest against the taller tree; calculate the length of the part of the taller tree which is above the point of contact.

6. (a) Simplify $\dfrac{\text{Sin } 30 \times \text{Sin } 45}{\text{Sin } 75}$

(b) Calculate the length of AB

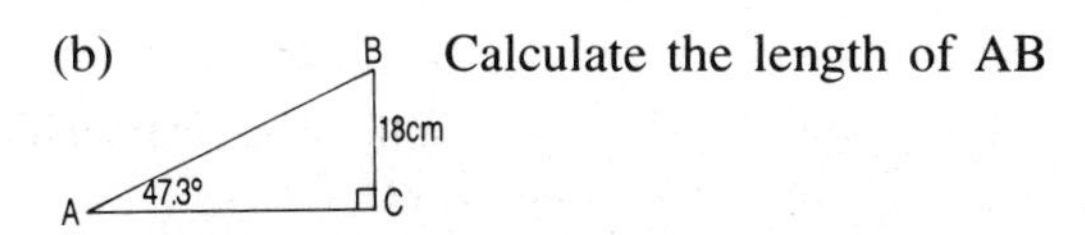

(c) Explain why the ratio of the side of a square to its diagonal is approximately 7 : 10.

(d) Given that Sin x = $^{12}/_{13}$, calculate Tan x and Cos x.
(You must make your method clear, and express your answers in fractions.)

PYTHAGORAS AND TRIGONOMETRY SUPPLEMENTARY QUESTIONS

1. 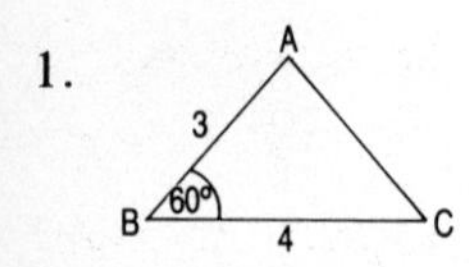

Calculate by using the Cosine rule the length of AC. Use your result and the Sin rule to calculate the size of angle C.

2. 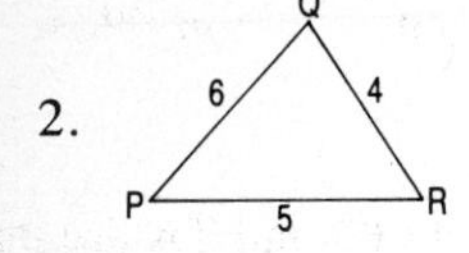

Use the Cosine rule to calculate the size of angle P.

3. 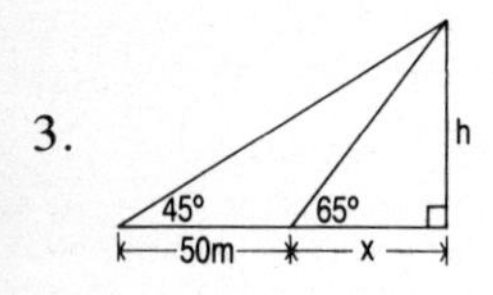

Calculate the length marked x and the length marked h.

4. If Sin x = $^{60}/_{61}$, calculate Tan x giving your answer as a fraction.

PROBABILITY AND STATISTICS

TIME 1 HOUR

1. For the set of numbers {8, 7, 2, 3, 1, 7, 7, 9}

   (a) Write down the mode and the median.

   (b) If another number is added to the list and the mean of all 9 numbers is 4 calculate the extra number.

   (c) What is the probability of choosing a prime number from the original set of numbers?

   (d) Calculate the probability that if two numbers are chosen from the original set that both are odd.

2. The table shows the results of a test for a group of children:

MARK	1	2	3	4	5	6	7	8	9	10
FREQ.	1	1	3	6	8	8	10	6	4	3

   (a) What was the median mark?

   (b) Calculate the mean mark.

   (c) If the pass mark was 7, what percentage of the children failed?

   (d) If the information were shown on a pie chart, what angle would represent the modal mark?

3. (i) A pupil asked 960 people which of the colours Red, Blue or Green they preferred; she presented the information on a pie chart and on a bar chart. The angle sectors for Blue and Red were 150° and 90° respectively.

   (a) How many people preferred Blue?

   (b) What percentage of the people said Green?

   (c) On the bar chart the height of the column representing Red was 7½cm; what height was the column for Blue?

   (ii) In a question about 2 dice, a pupil handed in the following correct answer "$1 - (5/6)^2 = 11/36$".

   (d) What might the question have been?

4. From a pack of 52 playing cards what is the probability of selecting

   (a) A red Queen or a diamond?

   (b) Two cards, neither of which is a club (no replacing)?

Two dice are thrown together; what are the chances that

   (c) both show the same number?

   (d) the difference between the two scores exceeds 3?

5. (a) The mean of 4 numbers is 15, and the mean of a further 6 numbers is 5; calculate the mean of all ten numbers.

   (b) The mean of 5 numbers is 14; when a sixth number is included the mean is unchanged. Calculate the sixth number.

   (c) A bag has 4 red buttons and 5 green ones. If two buttons are drawn out at random (without replacement), what are the chances that at least one is red?

   (d) Another bag has black (B) buttons and white (W) ones; Sally selected ten buttons at random, replacing the button each time. Her results were

   B, B, W, B, W, W, B, B, B, B

   The results reflect accurately the number of buttons in the bag. If there are 36 more black than white buttons, calculate how many buttons are in the bag.

6. (i) A bag contains 100 sweets; 30 are wrapped in red paper, 50 in blue, and the rest in white.

   (a) If a pupil choses a sweet at random calculate the probability that it has a white wrapper.

   (ii) Make a tree diagram to show the different ways of withdrawing 2 sweets (without replacement). From your diagram calculate the probability of choosing

   (b) 2 sweets in red wrappers

   (c) 2 sweets with the same wrappers

   (d) at least one sweet with a blue wrapper.

PROBABILITY AND STATISTICS SUPPLEMENTARY QUESTIONS

1. 3 coins are tossed; what are the chances of getting only two heads?

2. The marks, out of 50, for a group of candidates are shown in the table below:

MARKS	1 – 10	11 – 20	21 – 30	31 – 40	41 – 50
FREQ.	20	50	80	90	60

(a) Make a cumulative frequency for the information above.

(b) Choose scales of your own and draw the cumulative frequency curve.

(c) Estimate from your graph

(i) the median

(ii) the interquartile range.

(d) If Grade A was awarded to candidates who scored 76% or above, estimate how many candidates were awarded A's.

3. The chances of a certain imported plant surviving in England is 60%. Calculate the probability that if two plants are imported at most one will survive.

4. 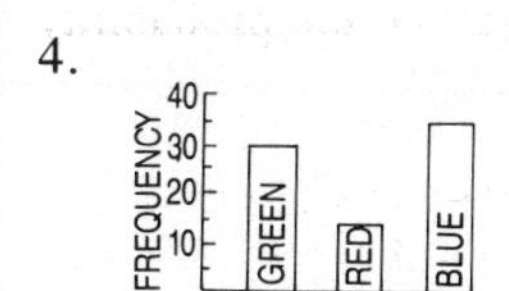

The bar chart shows the favourite colours of a group of children; represent this information on a pie chart of radius 3cm.

VECTORS

TIME 1 HOUR

1. $\underline{a} = \begin{pmatrix} 2 \\ 3 \end{pmatrix}$ and $\underline{b} = \begin{pmatrix} 1 \\ -4 \end{pmatrix}$

Find

(a) $2\underline{a} + \underline{b}$

(b) the modulus of $\underline{b}$, leaving your answer in root form.

(c) If also $\underline{c} = \begin{pmatrix} 9 \\ -3 \end{pmatrix}$, show that $\underline{a} + \underline{b}$ is parallel to $\underline{c}$.

(d) If $12\underline{a} + x\underline{b} = 4\underline{c}$, calculate x.

2. ABCD is a rhombus with $\overrightarrow{AD} = \underline{a}$, $\overrightarrow{AB} = \underline{b}$. E and F are the midpoints of DC and BC respectively.

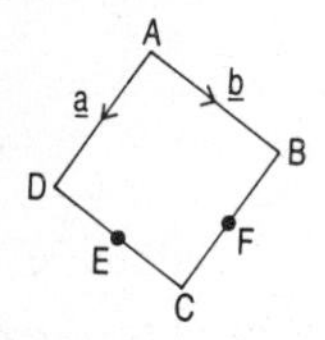

(a) Explain why $\underline{a}$ is not equal to $\underline{b}$.

(b) Write down in terms of $\underline{a}$ and $\underline{b}$

(i) $\overrightarrow{AC}$, and

(ii) $\overrightarrow{DB}$

(c) Prove that EF = ½DB

(d) Express the area of the triangle CFE as a fraction of the area of the whole rhombus.

3. ABC is a triangle with CD parallel to BA and equal to half of it. K is a point on AC such AK : KC is 2 : 1.

$\overrightarrow{BA} = \underline{a}$ and $\overrightarrow{BC} = \underline{b}$.

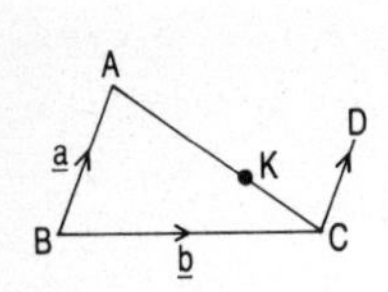

(a) Write down an expression for $\overrightarrow{AC}$ in terms of $\underline{a}$ and $\underline{b}$.

(b) Write down expressions for $\overrightarrow{AK}$ and $\overrightarrow{KC}$ in terms of $\underline{a}$ and $\underline{b}$.

(c) Find $\overrightarrow{BK}$ and $\overrightarrow{KD}$ in terms of $\underline{a}$ and $\underline{b}$, and simplify your expressions.

(d) Using your answers for (b) and (c) or otherwise show that B,K,D are collinear.

4. In $\triangle$ ABC, $\overrightarrow{AB} = \underline{m}$ and $\overrightarrow{AC} = \underline{n}$

D is a point on AB such that AD $= \frac{4}{7}$ AB;

E is a point on AC such that AE : EC = 4 : 3

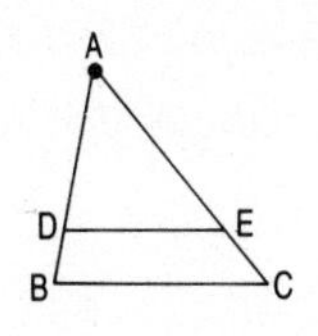

(a) Express $\overrightarrow{CB}$ in terms of $\underline{m}$ and $\underline{n}$.

(b) Express $\overrightarrow{DC}$ in terms of $\underline{m}$ and $\underline{n}$.

(c) Write down expressions for $\overrightarrow{AE}$ and $\overrightarrow{EC}$ in terms of $\underline{n}$.

(d) By expressing $\overrightarrow{DE}$ in terms of $\underline{m}$ and $\underline{n}$, show that $k\overrightarrow{DE} = \overrightarrow{BC}$ and state the value of k.

5. $\underline{a} = \begin{pmatrix} 4 \\ -1 \end{pmatrix}$ and $\underline{b}$ $\begin{pmatrix} -3 \\ x \end{pmatrix}$

(a) If $\underline{a} - \underline{b} = \begin{pmatrix} y \\ 7 \end{pmatrix}$ calculate x and y.

(b) What is the modulus of 3$\underline{a}$?

(c) On a co-ordinate grid with O as origin $\overrightarrow{OA} = \begin{pmatrix} 3 \\ 4 \end{pmatrix}$; write down three more position vectors, $\overrightarrow{OX}$, $\overrightarrow{OY}$ and $\overrightarrow{OZ}$, that have the same modulus as $\overrightarrow{OA}$ (use whole numbers only).

(d) If $\overrightarrow{OA} = \begin{pmatrix} 3 \\ 4 \end{pmatrix}$ and $\overrightarrow{AB} = \begin{pmatrix} -1 \\ 3 \end{pmatrix}$, what is $\overrightarrow{OB}$?

6. ABCDEF is a regular hexagon 'centre' O; $\overrightarrow{AB} = \underline{a}$ and $\overrightarrow{BC} = \underline{b}$.

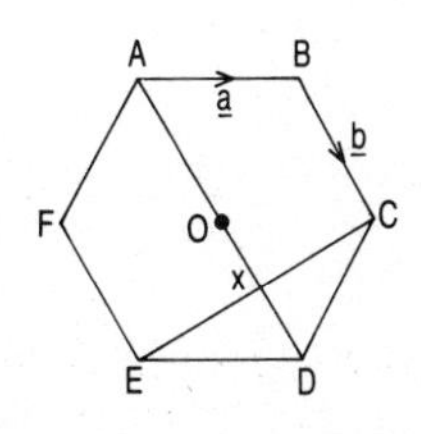

(a) Express $\overrightarrow{OB}$ in terms of $\underline{a}$ and $\underline{b}$.

(b) Express $\overrightarrow{FD}$ in terms of $\underline{a}$ and $\underline{b}$.

X is the point of intersection of AD and EC.

(c) Express $\overrightarrow{EC}$ in terms of $\underline{a}$ and $\underline{b}$.

(d) What is the ratio of AD : XD?

VECTORS

SUPPLEMENTARY QUESTIONS

1.

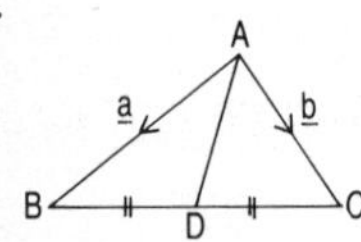

If $\vec{AB} = \underline{a}$ and $\vec{AC} = \underline{b}$, and D is the midpoint of BC, express in terms of $\underline{a}$ and $\underline{b}$
(i) $\vec{DB}$ (ii) $\vec{AD}$.

2. In the diagram $\vec{AB} = \underline{a}$ and $\vec{AD} = \underline{b}$; X is a point on the diagonal DB such that DX : XB is 4 : 1. XY is parallel to BC.

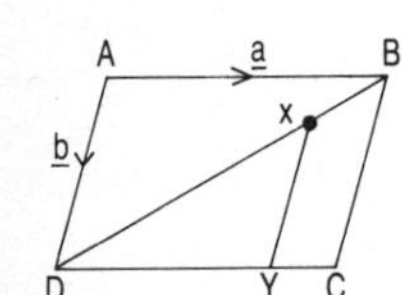

(a) Express $\vec{DX}$ in terms of $\underline{a}$ and $\underline{b}$.

(b) If $\vec{XY} = \frac{4}{5}\underline{b}$, deduce that ABCD is a parallelogram.

3.

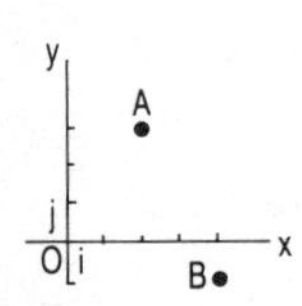

If $\vec{OA} = 2i + 3j$ and $\vec{OB} = 4i - j$, express $\vec{AB}$ in terms of i and j.

What is $|\vec{OA}|$?

4. If $\underline{a} = \begin{pmatrix} 8 \\ -9 \end{pmatrix}$ and $\underline{b} = \begin{pmatrix} -5 \\ 6 \end{pmatrix}$

Prove that $\underline{a} + \underline{b}$ is parallel to $\underline{c}$, where $\underline{c} = \begin{pmatrix} -2 \\ 2 \end{pmatrix}$

MIXED EXAMPLES 1

TIME 1½ HOURS (approximately)

1. The sketch represents the velocity/time graph of a particle over an 8-second period.

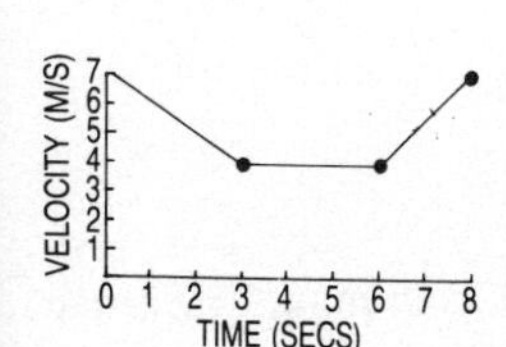

(i) What is the acceleration during the final 2 seconds?

(ii) Calculate the distance travelled by the particle during the 8 seconds.

2. Solve the equations

(a) $\dfrac{3 - x}{4} = 1 - x$

(b) $\dfrac{4}{x + 1} = \dfrac{3}{2x}$

3. Sketch 3 Venn diagrams to illustrate the following:

(i) $A \cap B = \phi$ (ii) $A \cap B = B$ (iii) $C \subset (A \cap B')$

4. (i) Simplify $(2\frac{1}{2} + 1\frac{3}{4}) \div 1\frac{1}{16}$

(ii) Find $\sqrt{1\frac{25}{144}}$

(iii) How many prime numbers are there between 60 and 70?

5. $A = \begin{pmatrix} 2 & 1 \\ 5 & 3 \end{pmatrix}$, $B = \begin{pmatrix} 2 \\ -1 \end{pmatrix}$.

Find (i) AB (ii) A^{-1} (iii) $\begin{pmatrix} x \\ y \end{pmatrix}$ such that $A \begin{pmatrix} x \\ y \end{pmatrix} = \begin{pmatrix} 2 \\ -1 \end{pmatrix}$

6. Solve the equation: $x^2 - 9x - 22 = 0$.

7. The volume of a sphere is $20cm^3$. Calculate its radius.

8.

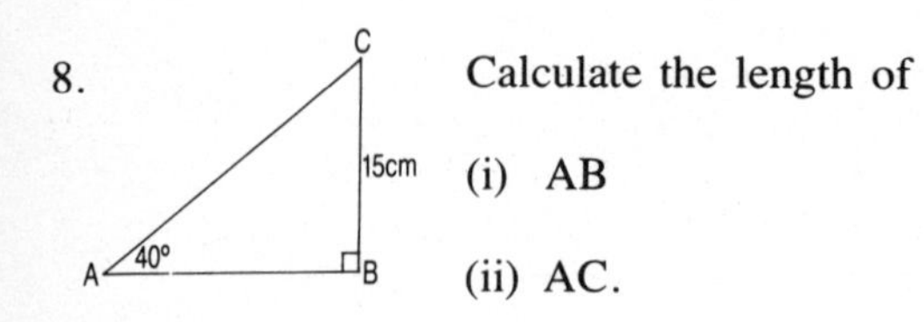

Calculate the length of

(i) AB

(ii) AC.

9. Taking £1 = 205 Spanish pesetas, convert £175 to ptas. The bank deducts 1% in commission on all currency transactions; Calculate how many pesetas a tourist will get for £175.

10. (i) Simplify: $3^\circ \times (2)^{-5}$

(ii) Express as a power of 2: $\dfrac{64^{1/3}}{512}$

11. The area of an equilateral triangle is $30cm^2$; Calculate the length of its side.

12. Prove that the co-ordinates $(-3, 2)$, $(1, -4)$, $(9, -16)$ are collinear.

13. A meal in a restaurant cost £18.40, including VAT at 15%. How much would the meal cost without VAT?

14. {1, 4, 3, 7, 8, 9, 1}.

(a) From this set of numbers, write down the median. Calculate the mean.

(b) If a number is selected at random, calculate the probability that it is a square number.

15.

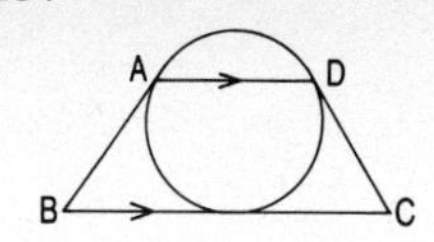

In the diagram BC, BA, and CD are tangents to the circle; the chord AD is parallel to BC.

Prove that AB = DC.

16.

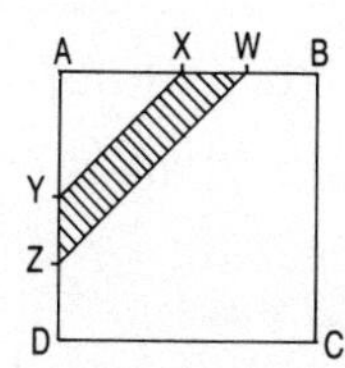

ABCD is a square. X and Y are the midpoints of AB and AD respectively;

AW = ¾ AB, AZ = ¾ AD.

What fraction of the square is shaded?

17.

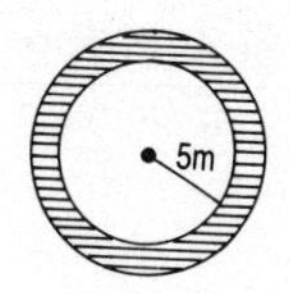

Two circles have the same centres; the radius of the inner circle is 5m and the area of the shaded border is 120m^2. Calculate the radius of the outer circle.

18.

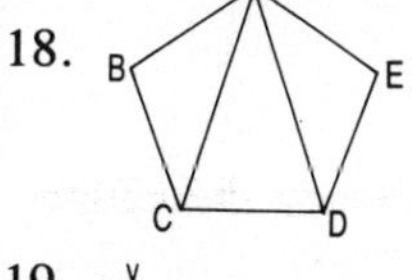

ABCDE is a regular pentagon;

Calculate the size of each angle in triangle ACD.

19.

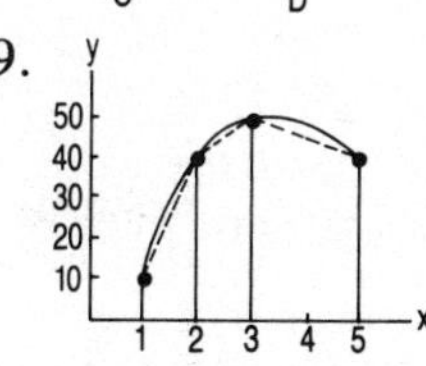

Estimate the area under the curve by using the given trapezia.

20. Write down an odd number (less than 100) that has only 5 factors.

21. For what values of x is:

(i) $3x - 1 > -10$?

(ii) $1 - 3x > -10$?

(iii) $-3 < 2x - 1 < 12$, where x is an integer?

22. $E = \{1, 2, 3, 4, 5, 6, 7, 8, 9\}$
$A = \{\text{primes}\}$; $B = \{\text{multiples of } 3\}$

Find

(i) A' (ii) $n(A \cap B')$ (iii) $A' \cap B'$

23. $f(x) = x^2 - 3$; find (i) $f(-2)$, (ii) a if $f(a) = 0$

24. Make a sketch of the graph of $y = \sin x$ for values of x from 0° to 90°, at 15° intervals. Use 5cm for 1 unit on the y-axis.

25. In the triangle ABC, DE is parallel to BC. AD = 3cm, DB = 5cm and AE = 4cm.

Calculate the length of EC.

If also BC = 12cm, find DE.

26. Which of the following are rational numbers:

(i) $\sqrt{3025}$ (ii) $\sqrt{2}$ (iii) π (iv) 0 (v) $(\frac{1}{2})^3$?

27.

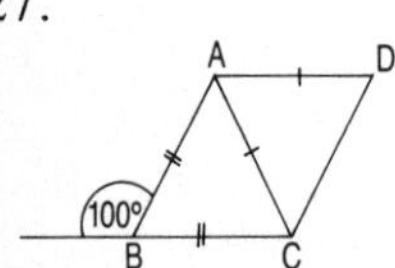

In triangle ABC, AB = BC, and the exterior angle at B = 100°. AD is drawn parallel to BC, and AD = AC.

Calculate angle D.

28. For how many years must £3,000 be invested at 5½% simple interest, in order to amount to £4,320?

29. What is the equation of the line which is parallel to $3y + 4x = 12$, and passes through $(-1, -1)$?

30. What is the modulus of the vector $\begin{pmatrix} -4 \\ 5 \end{pmatrix}$?

MIXED EXAMPLES Paper 2

NO TIME LIMIT

[Allow approximately 12½ minutes a question. Questions 11-15 may not all be on your syllabus.]

1. (a) y varies inversely as the square of x, and y is 1.2 when x is 5. Calculate x when y is 0.3.

   (b) Simplify: $\frac{6x - 3}{4} - \frac{5x - 3}{5}$

   (c) Find the equation of the straight line which passes through (1, − 3) and has a gradient of −2.

   (d) Use the coordinates below to draw the graph of $y = x^2 - x - 6$

x	−3	−2	−1	0	1	2	3	4
y	6	0	−4	−6	−6	−4	0	6

   Use your graph to solve the equations

   (i) $x^2 - x - 6 = 2$

   (ii) $x^2 - 2x - 4 = 0$

2. 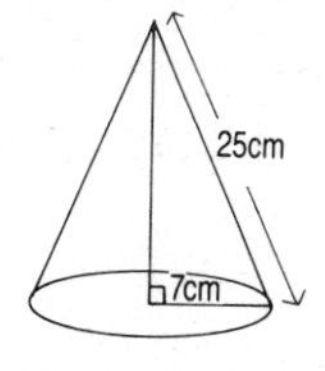

   (a) Calculate the height of the cone.

   (b) Calculate the volume of the cone, giving your answer to 3 significant figures.

   (c) Sketch the net of the cone, and calculate the sector angle. [You are advised not to substitute for π.]

   (d) Calculate the area of the remaining part of the circle from which the sector was cut?

3.

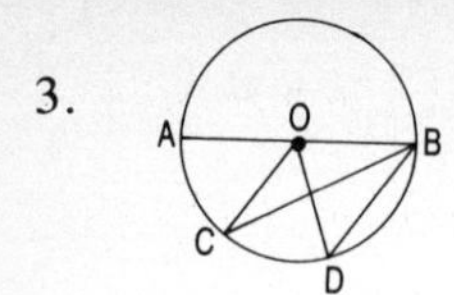

(i) In the diagram AOB is the diameter and angle OCB is 30°.

Calculate (a) angle AOC.

(b) angle DOB if BD is parallel to OC.

(ii) OA and OB are radii and angle ACB = 110°

(c) Calculate the size of the angle AOB.

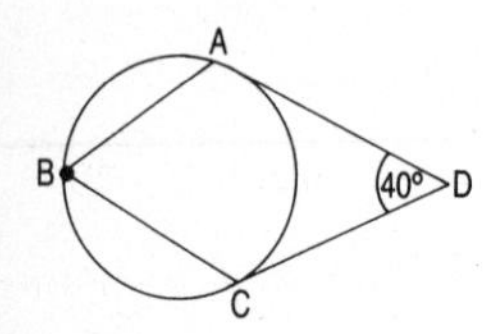

(iii) Given that DA and DC are tangents to the circle, and B is a point on the major arc AC, calculate the size of angle B.

4. Construct a triangle ABC, with AB = 8cm, AC = 5cm and BC = 6cm.

Find the locus of the points which are equidistant from AB and AC. Let the locus meet BC at E.

Draw through E the locus a moving point E_1, such that the area of $\triangle EAB$ = the area of $\triangle E_1AB$.

How many of these triangles will be right-angled?

5. Draw the triangle ABC on squared paper where A is (1, 4), B (3, 2) and C (4, 1).

(a) Reflect the triangle in the line y = x. State the matrix of this transformation.

(b) Apply the matrix $\begin{pmatrix} 1 & 0 \\ 0 & -1 \end{pmatrix}$ to triangle ABC; state the transformation which this matrix represents.

(c) What combination of transformations does the matrix $\begin{pmatrix} 0 & -2 \\ -2 & 0 \end{pmatrix}$ represent? (You may apply it to $\triangle ABC$ if you wish.)

(d) Another image of ABC is given by the co-ordinates P(−2, 5), Q(0, 3) and R(1, 2). Plot these co-ordinates and describe the transformation which maps $\triangle ABC$ on to $\triangle PQR$.

6. The results of a survey into people's ages were given as follows:

AGE	20–24	25–29	30–34	35–39	40–44	45–49
FREQ.	40	80	320	150	50	60

(a) What was the modal age group?

(b) Calculate the mean age of the group – stating any assumptions you make.

(c) In which age group will the median occur?

(d) If two people are chosen at random from the sample what is the probability that at least one will be 35 or over?

7.

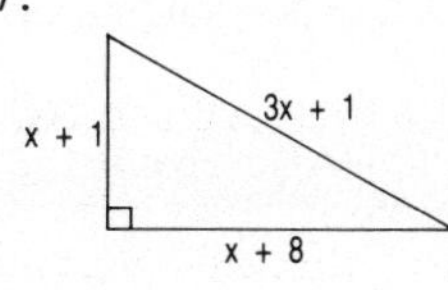

(a) The sides of a right angled triangle are $(x + 1)$cm, $(x + 8)$cm and $(3x + 1)$cm.

Use Pythagoras to calculate x [no credit will be given for an answer found by inspection].

(b) Given that AB = 5cm, $\hat{B} = 90°$ and Tan C = ¾ calculate the length of BC.

Find also the length of the perpendicular from B to AC.

8. (i) Let $\underline{r} = \begin{pmatrix} 2 \\ -3 \end{pmatrix}$ and $\underline{s} = \begin{pmatrix} 1 \\ -4 \end{pmatrix}$.

(a) If $k\,\underline{r} + p\,\underline{s} = \begin{pmatrix} 7 \\ -13 \end{pmatrix}$, find k and p.

(b) What is the modulus of $\underline{r} + \underline{s}$?

(ii)

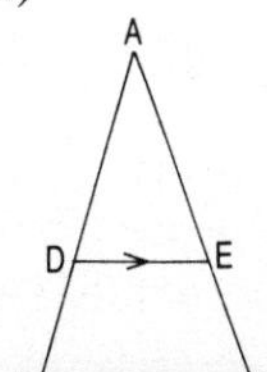

In $\triangle ABC$ let $\overrightarrow{AB} = \underline{a}$ $\overrightarrow{AC} = \underline{b}$.

(c) Write down $\overrightarrow{BC}$ in terms of $\underline{a}$ and $\underline{b}$.

If also DE is parallel to BC and DE = ⅔BC, find $\overrightarrow{BE}$ in terms of $\underline{a}$ and $\underline{b}$.

(d) What is the ratio of the areas of the two triangles in the diagram?

9. (a) Express 1728 as the product of prime powers. [Note: $12^3 = 1728$]

(b) An investment amounts to £360 after 4 years at 5% per annum. Calculate the original investment. [Hint: Let the original investment be £100x]

The value of a house appreciates by 5% each year. If it is now valued at £75 000, calculate, to the nearest £1,000,

(c) what the house was worth last year, and

(d) what it will be worth 3 years from now.

10. (i)

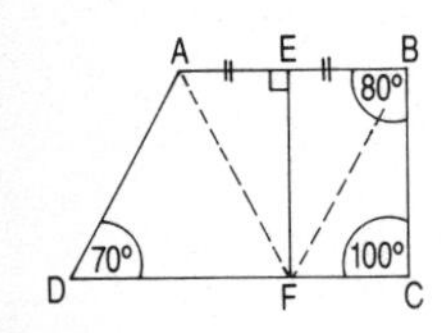

In the quadrilateral ABCD B = 80°, C = 100°, D = 70°.

E is the midpoint of AB and FE is perpendicular to AB.

(a) Calculate the size of angle DAB; what is the distinct name for ABCD?

(b) Prove that $\triangle^s$ AFE and BFE are congruent.

(ii) (c) The sum of the degrees of a polygon is 4500°; how many sides has the shape?

(d) Is it possible for a regular polygon to have each interior angle equal to 147½°? Explain your answer.

11. Given f(x) = 1 − 4x, and g(x) = 4x − 1.

(a) Find (i) f(3) (ii) gf(3) (iii) m if f(m) = 3.

(b) Express fg(x) in the form fg(x) ⟶ .

(c) What is the inverse of g(x)?

(d) For what value of x is f(x) = g(x)?

12. £600 is invested for 4 years at 8% per annum, Compound Interest. How much does it amount to, at the end of the period? Give your answer to the nearest £.

For how long would the £600 have to be invested at 8% per annum Simple Interest to yield the same amount?

13. The results of a mathematics test were grouped as below:

MARKS	1–20	21–40	41–60	61–80	81–100
FREQ.	10	20	30	25	15

(a) Make a cumulative frequency for the results.

(b) Using scales of your own draw the cumulative frequency graph.

(c) From your graph estimate

(i) the median

(ii) the interquartile range.

(d) If the top 20% of the candidates were awarded grade A, what mark did a pupil need in order to get A?

14.

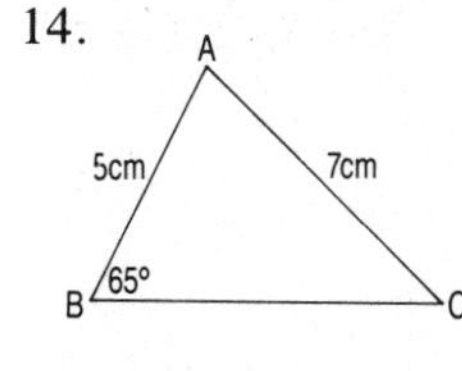

(a) Use the Sine rule to calculate the size of angle C.

(b) Calculate the length of BC.

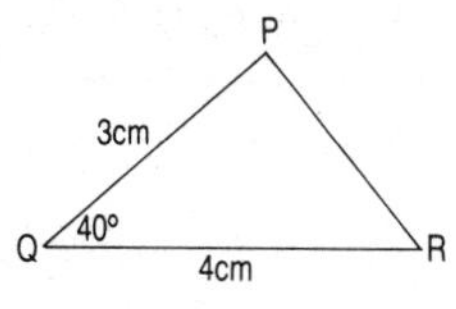

(c) Use the Cosine rule to calculate the length of PR.

(d) In a triangle LMN, LM = 8cm, MN = 5cm, LN = 6cm. Calculate the size of the smallest angle.

15. (a) Sketch Venn diagrams to illustrate the following relationships between the given sets:

(i) $A \cap B \cap C = \phi$ (ii) $(A \cup B) \subset C$

(b) When asked which games they liked 140 children gave the following replies: 70 said Hockey, 50 said Badminton, 40 said Rugby; 13 said Hockey and Badminton, 11 said Hockey and Rugby, 18 said Badminton and Rugby. 21 did not like any of the 3 games. Let x represent the number who said all 3; copy and complete the Venn diagram below; hence find x.

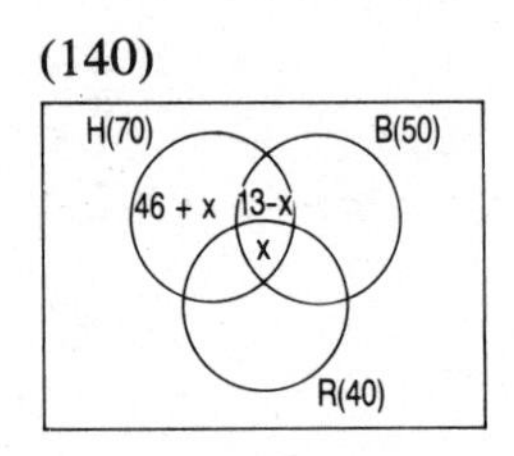

(c) Describe the group of 21 children using set notation.

PART IV

ANSWERS TO PRACTICE PAPERS CONTAINED IN PART III

ALGEBRA AND GRAPHS

1. (a) 18 (b) 6 (c) $x = 1, y = -1$ (d) $\frac{2}{3}$ or $1\frac{1}{2}$

2. (a) $^3/_2$ (b) $y = x - 1$ (c) Sketch (d) $y = x$; (1,1); (−1, −1)

3. (a) 8, 0, −4, 0 (b) graph (c) 4 (d) 2.82 or 0.18

4. (a) 3 (b) $p = \sqrt{\frac{t}{3}}$ (c) 50 (d) −40

5. (a) 64 (b) $21x - 19$ (c) $3x(x + 2y)$ (d) $x = \dfrac{n - m}{7}$

6. (a) 11 (b) 3 (c) 4 (d) $^1/_7$

SUPPLEMENTARY QUESTIONS

1. $(4x + 5y)(4x - 5y)$; $2\frac{1}{2}$

2. $p = \dfrac{q - qx}{bx - a}$

3. Area ≈ 20

4. $\pm ^4/_3$

AREA AND VOLUME

(Answers may vary slightly depending on the value of π chosen).

1. (a) 169.56 (b) 5.66 (c) πr^3 (d) 4

2. (a) 113.0 (b) 8:27 (c) 113.04 (d) 4:9

3. (a) $2x^2 + 40x$ (b) 8 (c) 48 (d) $1\frac{1}{2}$

4. (a) 301/302 (b) 297 (c) 60π (d) 1:16

5. (a) 6.18 (b) 4082 (c) 179.1° (d) 100.96

6. (a) 659.4 (b) 6.5 (c) $31\frac{1}{2}$ (d) 144, 112

SUPPLEMENTARY QUESTIONS

1. 2.29

2. 9,270

3. $h = 4r$

4. 4.89

CIRCLE GEOMETRY

1. (a) 145° (b) 35° (c) 72½° (d) 17½°

2. (a) 100° (b) 30° (c) $\hat{B} \neq \hat{D}$ (e.g.) (d) ABZ, XYZ, CDZ

3. (a) 80° (b) 40°, 25°, 115° (c) 5:13 (d) $108/5\pi$

4. (a) Proof (b) 60° (c) 110° (d) Proof

5. (a) 110° (b) 80° (c) 29° (d) 61°

6. (a) Proof (b) Proof (c) Proof (d) 10°

SUPPLEMENTARY QUESTIONS

1. 36° 2. 100° 3. 115° 4. 540°/7

CONSTRUCTIONS (INCLUDING BEARINGS & SCALES)

1. (ii) 1:320 000; 12½

2. (c) 34.5° (d) 2 hrs 51 mins

3. (a) 500m (b) 24cm (c) 180 + x, 180 + x, x − 180

5. (i) 36.5cm^2

SUPPLEMENTARY QUESTIONS

1. (a) 1:800 000 (b) 1⅛cm^2 (c) 60km

2. 27 hrs 51 mins.

4. 7cm

GEOMETRY OF RECTILINEAR FIGURES

1. (a) 40° (b) 2y − 180 (c) 9 (d) 1800

2. (a) 100° (b) Parallelogram (c) 120 (d) 8

3. (a) Proof (b) 60° (c) Rhombus or rectangle (d) Yes

4. (a) 135° (b) Proof (c) Proof (d) Proof

5. (a) 5 (b) 162° 3. 81 (d) Proof

6. (a) Reflection in perpendicular bisector of AB
   (b) Rotation 60° about 0
   (c) Translation 1 unit to the left
   (d) Rotation of 180° about the midpoint of OB

SUPPLEMENTARY QUESTIONS

1. 22 2. 180 − A 3. $\hat{D} = \hat{C} = 25°$ 4. Proof

MATRICES & TRANSFORMATIONS

1. (a) $\begin{pmatrix} -1 & -6 \\ 1 & 2 \end{pmatrix}$ (b) ⅓ $\begin{pmatrix} 1 & 3 \\ -1 & 0 \end{pmatrix}$

   (c) $\begin{pmatrix} 1 & 0 \\ 1 & 0 \end{pmatrix}$ (d) ⅓ $\begin{pmatrix} -1 & -6 \\ 1 & 2 \end{pmatrix}$

2. (a) Diagram (b) Reflection in y = −x

   (c) $\begin{pmatrix} 1 & 0 \\ 0 & -1 \end{pmatrix}$ (d) Reflection in y = x

3. (a) $\begin{pmatrix} \frac{1}{2} \\ \frac{1}{2} \end{pmatrix}$; $\begin{pmatrix} 75 \\ 55 \end{pmatrix}$ (b) $\begin{pmatrix} 1 & 0 \\ 0 & 1\frac{1}{5} \end{pmatrix}$; $\begin{pmatrix} 80 & 70 \\ 60 & 72 \end{pmatrix}$

(c) $\begin{pmatrix} -1 \\ 2 \\ 2 \end{pmatrix}$ (d) $\begin{pmatrix} \frac{1}{2} & \frac{1}{2} \\ \frac{1}{2} & \frac{1}{2} \end{pmatrix}$

4. (a) $\begin{pmatrix} 0 & 1 \\ 1 & 0 \end{pmatrix}$ (b) Reflection in y − axis (c) Ref in y = x

(d) $\begin{pmatrix} 0 & 1 \\ 1 & 0 \end{pmatrix}$

5. (a) 1 (b) Rot 36.9° about (0,0) (c) trans $\begin{pmatrix} -4 \\ 0 \end{pmatrix}$

followed by a 90° Rot about (−1,2) (e.g.)

6. (a) (0, 2) (b) n = 0, x = −3 (c) $\begin{pmatrix} 6 & 2d \\ -3 & -d \end{pmatrix}$

SUPPLEMENTARY QUESTIONS

1. Enlargement scale factor $2\sqrt{2}$ centre (0,0) followed by a rotation 315° about (0,0)

2. a = b = 2

3. Rotation of 180° about (0,0)

4. Reflection in y = −x

NUMBERS AND FINANCE

1. (a) £10.40 (b) £8 (c) 84p (d) 63p
2. (a) 31, 37 (b) 12^2 (c) $2 \times 5 \times 31$ (d) $2^6 \times 5^2$
3. (a) 6 (e.g.) (b) 2^9 (c) 4.8×10^3 (d) 1.4×10^{-8}
4. (a) £975 (b) £2000 (c) 240; £15
5. (a) 1.45pm (b) 190 mins (c) 7 (d) 19%
6. (a) 40 (b) 49500, 50499 (c) £4500 (d) $^1/_{50}$

SUPPLEMENTARY QUESTIONS

1. £182; 8.7% 2. 15%; £64000 3. 1×4^{-10} 4. 36%

PYTHAGORAS & TRIGONOMETRY

1. (a) 10cm (b) 73.7° (c) 7cm (d) 69.2
2. (a) 24 (b) $\sqrt{288}$ (c) 73.7° (d) 61.3°
3. (a) 34.7 (b) 46.1° (c) 21.1 (d) 124.7°
4. (a) 1:15 (b) 3.8° (c) 9750m (d) 3.5°
5. (a) 28.7° (b) 8.34 (c) 20.6° (d) 53.1cm
6. (a) 0.366 (b) 24.5 (c) e.g. $\text{Sin } 45° = \frac{1}{\sqrt{2}} \approx 0.7$

   (d) $^{12}/_5$, $^5/_{13}$

SUPPLEMENTARY QUESTIONS

1. $\sqrt{13}$; 46.1° 2. 41.4° 3. x = 43.7, h = 93.7 4. $^{60}/_{11}$

PROBABILITY AND STATISTICS

1. (a) 7, 7 (b) -8 (c) $\frac{5}{8}$ (d) $\frac{15}{28}$

2. (a) 6 (b) 6.16 (c) 54% (d) 72°

3. (a) 400 (b) $33\frac{1}{3}$% (c) $12\frac{1}{2}$cm (d) .... at least one 5.... (e.g.)

4. (a) $\frac{7}{26}$ (b) $\frac{19}{34}$ (c) $\frac{1}{6}$ (d) $\frac{1}{6}$

5. (a) 9 (b) 14 (c) $\frac{13}{18}$ (d) 90

6. (a) $\frac{1}{5}$ (b) $\frac{29}{330}$ (c) $\frac{37}{99}$ (d) $\frac{25}{33}$

SUPPLEMENTARY QUESTIONS

1. $\frac{3}{8}$

2. (a) 20, 70, 150, 240, 300 (b) graph
   (c) (i) 30 (ii) 39 − 21 = 18 (d) 90

3. 64%

4. Angles 135°, $67\frac{1}{2}$°, $157\frac{1}{2}$°

VECTORS

1. (a) $\begin{pmatrix} 5 \\ 2 \end{pmatrix}$ (b) $\sqrt{17}$ (c) Proof (d) 12

2. (b) $a + b$; $b - a$ (c) Proof (d) $\frac{1}{8}$

3. (a) $b - a$ (b) $\frac{2}{3}(b - a)$ (c) $\frac{1}{3}a + \frac{2}{3}b$; $\frac{1}{6}a + \frac{1}{3}b$

4. (a) $m - n$ (b) $n - \frac{4}{7}m$ (c) $\frac{4}{7}n$; $\frac{3}{7}n$ (d) $k = \frac{7}{4}$

5. (a) $x = -8$, $y = 7$ (b) $\sqrt{153}$ (c) e.g. $\begin{pmatrix} 0 \\ 5 \end{pmatrix}$; $\begin{pmatrix} 5 \\ 0 \end{pmatrix}$; $\begin{pmatrix} 3 \\ -4 \end{pmatrix}$
   (d) $\begin{pmatrix} 2 \\ 7 \end{pmatrix}$

6. (a) $a - b$ (b) $a + b$ (c) $2a - b$ (d) 4:1

SUPPLEMENTARY QUESTIONS

1. (i) $\frac{1}{2}(a - b)$ (ii) $\frac{1}{2}(a + b)$

2. (a) $\frac{4}{5}(a - b)$; (b) Proof

3. $2i - 4j$; $\sqrt{13}$

MIXED EXAMPLES 1

1. (i) $1\frac{1}{2}$ (ii) $39\frac{1}{2}$ 2. (a) $\frac{1}{3}$ (b) $\frac{3}{5}$ 3. Diagrams

4. (i) 4 (ii) $1\frac{1}{12}$ (iii) 2 5. (i) $\begin{pmatrix} 3 \\ 7 \end{pmatrix}$ (ii) $\begin{pmatrix} 3 & -1 \\ -5 & 2 \end{pmatrix}$

(iii) $x = 7$, $y = -12$

6. 11 or -2 7. 1.68 8. (i) 17.9 (ii) 23.3

9. 35875, 35516.25 10. (i) $\frac{1}{32}$ (ii) 2^{-7}

11. 8.32 12. Proof 13. £16 14. (a) 4, $4\frac{5}{7}$ (b) $\frac{4}{7}$

15. Proof 16. $\frac{5}{32}$ 17. 7.95 18. 36, 72, 72

19. 160 20. 81 21. (i) $x > -3$ (ii) $x < 3$ (iii) 0, 1, 2, 3, 4, 5, 6

22. (i) $\{1, 4, 6, 8, 9\}$ (ii) 3 (iii) $\{1, 4, 8\}$

23. (i) -1 (ii) $\pm\sqrt{3}$

25. $6\frac{2}{3}$, $4\frac{1}{2}$ 26. (i), (iv), (v) 27. 65° 28. 8 years

29. $3y + 4x = -7$ 30. $\sqrt{41}$

MIXED EXAMPLES 2

1. (a) ± 10 (b) $\frac{10x - 3}{20}$ (c) $2x + y + 1 = 0$

   (d) (i) 3.37, -2.37 (ii) 3.24, -1.24

2. (a) 24 (b) 1230 (c) 100.8° (d) 1414

3. (a) 60° (b) 60° (c) 140° (d) 70°

4. 4

5. (a) $\begin{pmatrix} 0 & 1 \\ 1 & 0 \end{pmatrix}$ (b) Ref. in x − axis

   (c) Ref. in $y = -x$ fall by enl. scale factor 2, centre (0,0) or or vice versa (d) translation $\begin{pmatrix} -3 \\ 1 \end{pmatrix}$

6. (a) 30 − 34 (b) 33.9 (c) 30 − 34 (d) 0.605

7. (a) 4 (b) 6⅔; 4

8. (a) $k = 3, p = 1$ (b) $\sqrt{58}$ (c) $b - a$; $\frac{2}{3} b - a$ (d) 4:9

9. (a) $2^6 \times 3^3$ (b) £300 (c) £71000 (d) £87000

10. (a) 110° (b) Proof (c) 27 (d) No

11. (a) $-11, -45, -\frac{1}{2}$ (b) $5 - 16x$ (c) $\frac{x + 1}{4}$ (d) ¼

12. £816; 6.76

13. (a) 10, 30, 60, 85, 100 (c) 53; 74 − 36 = 38 (d) 78

14. (a) 40.3° (b) 7.45 (c) 2.57 (d) 38.6°

15. (a) Diagrams (b) $x = 1$ (c) $n(A' \cap B' \cap C')$ e.g.

Advanced Level Mathematics

R C Solomon

596pp £6.95 1988

This book is designed as **support** for formal tuition, rather than as a replacement for it. Relevant theorems are quoted, but they are not proved. Formulae are given, but they are not derived. Emphasis is on **worked** examples, **graded** exercises, examination questions and **common errors.** A typical Advanced Level syllabus contains a core of Pure topics, Mechanics and Statistics. This book covers all the material needed for such a syllabus.

Contents

Review Extracts

'The common error feature is very useful and clearly is the result of many years of teaching experience ... The principal use of the book would appear to be as a source of exercises and examination questions for students revising or repeating their A Level course.'

Theta

National Curriculum 11-14 Year Olds

Mathematics Attainment Tests Key Stage 3

S Burndred

128pp (approx) £3.50 July 1990

This book enables a pupil to know what level he or she has reached in mathematics in relation to the standards laid down within the National Curriculum.

It is expected to be of value to all pupils aged between 11 and 14.

National Curriculum - Key Stage 3 (11-14 year olds)

Average pupils aged 11 are expected to be capable of achieving Level 4 in the specified attainment targets. In their subsequent three years, they are expected to have moved on to achieve Level 6. High flyers may be able to achieve Level 8
It is therefore useful for all 11-14 year olds to know where they are at the moment, in order to appreciate what is expected of them in the National Curriculum.

Approach and Content

The book is grouped by the topics for which Attainment Targets and Programmes of Study have been laid down ie. Number, Algebra, Measures, Shape and Space, handling Data and Using and Applying Maths.

Within each topic, the mathematical knowledge required and Attainment Tests (half hour) applicable to each of the Levels 4, 5, 6 and 7 are given. In this way, for any topic the reader can see exactly what level they are at. Answers, with a marking scheme, are given in an Appendix.

There are also 'Mock Examinations' for Levels 5 and 6, which cover all the topics, also with answers.

GCSE Mathematics (Higher and Intermediate)

One Hour Practice Papers

J J McCarthy

(Higher) 96pp £2.50 1988
(Intermediate) 80pp £2.50 1989

Teachers, parents and pupils want to know what the pupils' chances are of passing at the level **entered**. What better way of fining out than to set aside an hour every few weeks throughout the course of study and test?

There are **two** sets of one hour Practice papers for **each** topic or group of topics (eg area and volume, etc.). One set can be used during the school term, and one for final revision. in all there are **ten** pairs of topic-based papers plus **two** pairs of mixed topics (to provide a format as close to examination requirements as possible).

Note

The GCSE Boards' Syllabuses, Sample papers and Examination papers have been closely studied and their main requirements incorporated. One hour was chosen as feasible in a double period at school and as a reasonable time to set aside for home self-testing.